DECISION-MAKERS FIELD CONFERENCE 2005

MINING IN NEW MEXICO

The Environment, Water, Economics, and Sustainable Development

L. Greer Price, Douglas Bland,
Virginia T. McLemore, and
James M. Barker, Editors

New Mexico Bureau of Geology and Mineral Resources
A Division of New Mexico Institute of Mining and Technology
2005

Mining in New Mexico: The Environment, Water, Economics, and Sustainable Development
L. Greer Price, Douglas Bland, Virginia T. McLemore, and James M. Barker, Editors

Copyright © 2005
New Mexico Bureau of Geology and Mineral Resources
Peter A. Scholle, *Director and State Geologist*

a division of

New Mexico Institute of Mining and Technology
Daniel H. López, *President*

BOARD OF REGENTS

Ex-Officio
Bill Richardson, Governor of New Mexico
Veronica C. Garcia, Secretary of Education

Appointed
Jerry A. Armijo, 2003–2009, Socorro
Richard N. Carpenter, 2003–2009, Santa Fe
Ann Murphy Daily, 1999-2011, Santa Fe
Sidney M. Gutierrez, 2001–2007, Albuquerque
Michaella J. Gorospe, 2005–2006, Socorro

DESIGN & LAYOUT: Gina D'Ambrosio
SERIES DESIGN: Christina Watkins

EDITING: Jane C. Love and Nancy S. Gilson
SERIES EDITOR: L. Greer Price

CARTOGRAPHY & GRAPHICS: Leo Gabaldon and Tom Kaus
CARTOGRAPHIC SUPPORT: Kathryn Glesener and Glen Jones
TABLES: Nancy S. Gilson

EDITORIAL ASSISTANCE: James Tabinski

Special thanks to
New Mexico Energy, Minerals and Natural Resources Department
Joanna Prukop, Secretary
and the Mining and Minerals Division
Bill Brancard, Director
for support of the printing of this year's guidebook.

New Mexico Bureau of Geology and Mineral Resources
801 Leroy Place
Socorro, NM 87801-4796
(505) 835-5420
http://geoinfo.nmt.edu

ISBN 1-883905-22-2
First Printing May 2005

COVER PHOTO: Sangre de Cristo Mountains near Taos
Copyright © Ralph Lee Hopkins

**The New Mexico Bureau of Geology and Mineral Resources
wishes to thank the following for their support
of this year's conference and guidebook:**

SUPPORTING AGENCIES

New Mexico Energy, Minerals and Natural Resources
 Department (NMEMNRD)
New Mexico Environment Department
New Mexico Institute of Mining and Technology
Los Alamos National Laboratory
U.S. Bureau of Land Management
New Mexico State Land Office

EVENT SPONSORS

American Institute of Professional Geologists
Amigos Bravos
Molycorp, Inc.
New Mexico Geological Society
Phelps Dodge Mining Company
Taos Gravel Products
Vulcan Materials, Inc.

PRINCIPAL ORGANIZERS

Douglas Bland, Technical Program Chair and
 Field Trip Leader
Virginia McLemore, Technical Program
John Pfeil, Technical Program
James Barker, Technical Program
Gretchen Hoffman, Technical Program
L. Greer Price, Guidebook
Paul Bauer, Program Coordinator
Susan Welch, Logistics Coordinator
Susie Ulbricht, Administrative Assistance

LOGISTICS

James Barker
Ruben Crespin
Gretchen Hoffman
Mark Mansell
Geoffrey Rawling
Stacy Wagner

PLANNING COMMITTEE

Ernie Atencio, Taos Land Trust
John Bemis, New Mexico State Land Office
Consuelo Bokum, 1000 Friends of New Mexico
Mike Bowen, New Mexico Mining Association
Bill Brancard, Mining and Minerals Division, NMEMNRD
Senator Carlos Cisneros, New Mexico State Legislature
Teresa Conner, Conner & Associates
Bill Dalness, Bureau of Land Management
Dale Doremus, Ground Water Quality Bureau, New
 Mexico Environment Department
Ron Gardiner, New Mexico State Legislature
Callie Gnatkowski, Office of U.S. Senator Pete Domenici
Ricardo Gonzales, New Mexico State Legislature
Steve Harris, Rio Grande Restoration
Mike Inglis, Earth Data Analysis Center, University of
 New Mexico
Gary King, New Mexico Mining Commission
Mike Linden, USDA Forest Service
Mary Ann Menetrey, Mining Environmental Compliance
 Section, New Mexico Environment Department
Charlie Nylander, Los Alamos National Laboratory
John Romero, New Mexico Office of the State Engineer
Peter Scholle, New Mexico Bureau of Geology and
 Mineral Resources
Brian Shields, Amigos Bravos
John Shomaker, John Shomaker & Associates, Inc.
Gary Stephens, Bureau of Land Management
Representative Mimi Stewart, New Mexico State
 Legislature
Ron Thibedeau, USDA Forest Service
Tony Trujillo, Phelps Dodge Mining Company
Rod Ventura, New Mexico Environmental Law Center
Allen Vigil, Taos County
Anne Wagner, Molycorp, Inc.

Contents

Preface... **vii**

An Introduction from the State Geologist – *Peter A. Scholle*... **1**

CHAPTER ONE
THE PHYSICAL AND HISTORICAL FRAMEWORK

The Geology and Landscape of the Taos Region –
  Paul Bauer... **7**
Hydrology and Water Supply in the Taos Region –
  John W. Shomaker and Peggy Johnson... **12**
A History of Agricultural Water Development in the Taos Valley –
  Peggy Johnson... **16**
Taos County's Mineral Heritage – *Robert W. Eveleth*... **21**
The History and Operating Practices of Molycorp's Questa Mine –
  Anne Wagner... **26**
Metal Deposits in New Mexico—How and Where They Are –
  Robert M. North and Virginia T. McLemore... **29**
Industrial Minerals in New Mexico –
  Peter Harben... **34**

CHAPTER TWO
ENVIRONMENTAL AND WATER QUALITY ISSUES

Public Perspective on Mining –
  Stephen D'Esposito... **41**
The Environmental Legacy of Mining in New Mexico –
  James R. Kuipers... **46**
Addressing Abandoned Mine Lands Issues in New Mexico... **50**
Watershed Protection and Restoration in New Mexico –
  David W. Hogge and Michael W. Coleman... **55**
Acid Rock Drainage – *Kathy Smith*... **59**
A River on the Edge – *D. Kirk Nordstrom*... **64**
Water Quality Regulation of Mining Operations in New Mexico –
  Mary Ann Menetrey... **68**
Engineering Challenges Related to Mining and Reclamation –
  Terence Foreback... **73**
Environmental Impacts of Aggregate Production – *William H. Langer,
  L. Greer Price, and James M. Barker*... **78**

CHAPTER THREE
POLICY, ECONOMICS, AND THE REGULATORY FRAMEWORK

An Industry Perspective on Mining in New Mexico –
  Patrick S. Freeman... **85**
Economic Impact of Mining on New Mexico – *John Pfeil*... **90**
The Economic Anomaly of Mining – *Thomas Michael Power*... **96**
Significant Metal Deposits in New Mexico—Resources and Reserves –
  Virginia T. McLemore... **100**
An Overview of the Regulatory Framework for Mining in New Mexico –
  Douglas Bland... **105**
The New Mexico Mining Act – *William Brancard*... **112**

Planning for Mine Closure, Reclamation, and Self-Sustaining Ecosystems
under the New Mexico Mining Act – *Karen Garcia and
Holland Shepherd*... **117**
Financial Assurance and Bonding – *Warren McCullough*... **121**
Financial Assurance for Hard Rock Mining in New Mexico –
Ned Hall... **125**
Will There Be Water to Support Mining's Future in New Mexico? –
John W. Shomaker... **128**

CHAPTER FOUR
SUSTAINABLE DEVELOPMENT, TECHNOLOGY, AND A LOOK TO THE FUTURE

Sustainable Development and Mining Communities –
Dirk van Zyl... **133**
Sustainable Aggregate Resource Management – *William H. Langer*... **137**
New Research and Technology for Mining and Mine Clean-up –
R. David Williams... **141**
How Science Can Aid in the Decision-Making Process –
Peter A. Scholle... **145**
Finding Solutions – *Julia Hosford Barnes and Mary Uhl*... **149**
The Role of Non-Government Organizations – *Brian Shields*... **152**
The Future of Mining in New Mexico – *Editors*... **156**

List of Contributors... **160**

Photo Credits... **166**

Acronyms... **167**

Generalized Stratigraphy of the Taos Region... **168**

Generalized Geologic Map of the Taos Region... **Inside back cover**

Preface

This is the fourth volume we've produced in conjunction with our Decision-Makers Field Conferences. These conferences are designed to provide decision makers with an overview of earth science and related policy issues of interest and importance to all New Mexicans. We produced the first volume in 2001 on *Water, Watersheds, and Land Use in New Mexico*. This was followed by volumes on *New Mexico's Energy, Present and Future* in 2002 and *Water Resources of the Lower Pecos Region* in 2003.

The conferences have been a resounding success. One of the highlights has been the guidebooks, which have taken on a life and significance that extends well beyond the conferences themselves. Designed to provide a broad overview of the topic at hand, with a focus on how science can aid in the decision-making process, these books have become references for decision makers, policy makers, industry, the environmental community, and the general public. Written for a non-technical audience, the books are intended to provide information that is otherwise not easily found, in an accessible style and format.

This year we tackled the topic of *Mining in New Mexico*, with a focus on the Taos region. But, as in previous years, the issues at hand are statewide issues. In particular we wanted to address those mining-related issues that face all of us in the years ahead: environmental and social concerns, policy and economics, regulation, and the issue of sustainability. We've tried to provide a balanced view rather than a comprehensive one, and if we've not provided answers to all of the questions, perhaps we have at least provoked significant thought and discussion.

We tried hard to focus on science, as always, because science and policy are (and should be) closely linked, particularly with regard to mining. So this year's volume includes a little more in the way of policy and regulation. Our authors were chosen for their ability to address these topics broadly, and with authority, based on their expertise and experience. We asked authors to rely on fact rather than opinion, but the papers invariably reflect to some degree the views of their authors. Those views do not necessarily represent the voice of the New Mexico Bureau of Geology and Mineral Resources or our partner agencies. This year's contributors are listed in the back of the volume, along with information about who they are and what they do. The contributors themselves are an important resource; many of them will remain involved with shaping the future of mining in New Mexico.

Whatever that future may hold, it will require tough—and informed—decisions on the part of decision makers, industry, and the general public. It is our hope that this compilation will go far toward helping us all make informed decisions, based on an understanding of the science, as well as the social, policy, and economic issues, that are involved. Our economic health, our environmental well-being, and the quality of life that we have come to take for granted in New Mexico all depend upon it.

—*The Editors*

Map of field trip area.

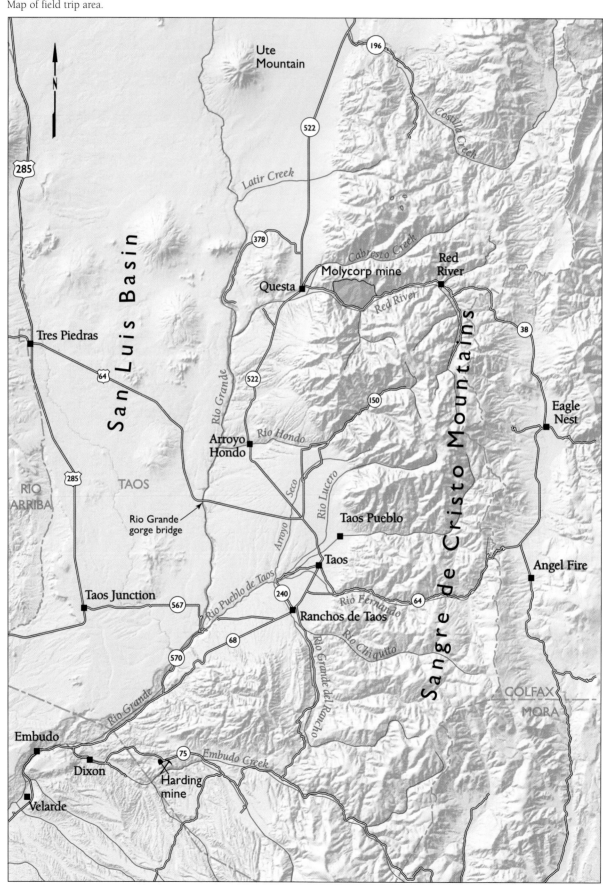

An Introduction from the State Geologist

Peter A. Scholle, *New Mexico Bureau of Geology and Mineral Resources*

Welcome to this decision-maker field conference, the fourth in our ongoing series of meetings dealing with geological and hydrological issues in New Mexico. These conferences—and this volume—are designed to provide New Mexico decision makers with the opportunity to see first-hand the influences and impacts of natural phenomena and human actions on our resources and landscapes. The conferences also provide an opportunity for participants to hear, see, and interact with leading scientific and technical experts from a wide range of partner organizations who will present material essential for an understanding of the relevant issues and their potential solutions. Those experts are the source of most of the papers in this volume. We strive to present a balanced program and to educate rather than lobby for specific legislation. Along with our many partners, however, we also hope that the information presented, the contacts made, the discussions engaged in, and a future of continued interactions after the trip all will lead to useful future legislation for New Mexico.

This year's meeting on mining-related issues deals with some of the most difficult and contentious topics we have tackled. Because mining issues revolve more around emotion-charged conflicts between differing societal values than around scientific disagreements, it is difficult to be dispassionate, and difficult to focus primarily on science. The social issues center on the differences between economic and environmental values. Clearly our society needs mining—it adds substantially to the economic welfare of the state and the nation (jobs, taxes, royalty revenues) and provides vital materials needed for our industrial economy. None of us, no matter how spartan our lifestyle, makes do without mined metals or industrial minerals. Such materials are found in our foods, our vehicles, our houses and appliances, our office buildings, our roads, and virtually everything else we use.

At the same time, however, we all want a clean and healthy environment, with safe water supplies, clean air to breathe, and pristine scenery to enjoy. Mining by its very nature impinges on some of those environmental values, at least during the working life of a mine.

This dichotomy between economic and environmental values is at the heart of the conflicts over regulation, permitting, and cleanup in the mining arena. So where does science fit into such values conflicts? The young field of environmental science has clearly increased public concerns about ecological and health-related issues. Epidemiological studies on health effects of pollutants, long-term air- and water-quality monitoring studies, ecosystems investigations, and other scientific investigations have defined the hazards associated with a wide variety of industrial activities, including mining. But science does not really inform the debate, as long as the debate is focused on the relative merits of two sets of social priorities. The question of which is more important, generating economic benefits or protecting the environment, is not one for which science can provide an answer. But science can provide background information to make that decision more rational and less emotional. How much ore is present in a prospect area, how much will it cost to extract, how long can the deposit be mined at certain rates, how much revenue will it add to state or local coffers—these are questions that science can help to answer. On the environmental side, how much air or water pollution will be generated, how can pollution most effectively and economically be minimized, what substitutes can be found for especially scarce or polluting materials are all questions for which science and technology can offer at least partial answers. Equally significant, it is the growth of science, especially environmental and medical sciences, that has also led to a profound shift in societal perceptions in the economic-environmental values discussion.

Companies, including those in the extractive industries, work to societal standards of their time, and those standards have changed dramatically in the past century. I happen to collect postcards of the oil industry and so will illustrate this point with several such images, but the same point could also be made using illustrations from the mining industry. Images on postcards generally reflect things of which we are proud. Today beautiful scenery, wildlife, modern buildings or monuments, and similar themes grace our postcards. It is not easy to find modern cards depicting oil wells, refineries, storage tanks, or other symbols of the petroleum industry, even in areas that have been or now are major oil producing regions. The same is true for mines, smelters, and other factories.

Postcards of the late nineteenth and early twentieth centuries, however, were very different in this regard. Thousands of towns across the nation, from Pennsylvania to California, issued postcards depicting oil fields. The scenes are perhaps horrific to the modern, environmentally conscious eye: forests of wooden derricks on deforested hillslopes, open lakes of oil, gushers and torpedoed wells spouting oil skyward, flaming storage tanks and oil fields. But the scenes also commonly show proud citizens in their finest clothing, walking under parasols through a rain of black gold or gazing at burning tanks. Identical scenes of environmental devastation are commonly claimed by multiple towns, each vying to become the industrial hub of their region. The inscriptions on the cards proclaim the manifest destiny of this nation, laud oil exploitation as a key part of the industrial revolution, and use the scenes to draw new population to the region. "Wish you were here" is a common, and sincere, sentiment expressed in the writings on the cards.

The point to be made here is that it is not really fair to hold modern industry solely responsible for the sins of earlier days. There is indeed a legacy of spoiled landscapes, air and water pollution, collapsing mine shafts, and other problems associated with past min-

A postcard depicting a burning oil storage tank in western Ohio from around 1910. Note the well-dressed citizens who have come out to have their group portrait taken in front of the billowing plume of black smoke.

A postcard view of a raging oil field fire from around 1900. Such fires were common occurrences given forests of wooden derricks that virtually touched at their bases, uncovered rivers and lakes of oil (foreground), and the common use of open blacksmith fires for dressing the cable tools used to drill these wells. This postcard is from Baku in the Russian Empire (now Azerbaijan), but comparable scenes were common in U.S. oil fields of the time.

ing, in New Mexico and throughout the nation. But in earlier days there was no widespread recognition of the potential problems, no scientific information on the health effects of pollutants. Instead, there was widespread jubilation at the jobs and wealth that mining and other extractive industries brought to the economy of a struggling and growing nation.

So how do we move forward more intelligently with mining today and in the future? We cannot expect companies working in a free-market economy, whether in the nineteenth century or today, to substantially exceed mandated environmental standards. It is those scientifically established and legislatively mandated standards that set the benchmark those companies must achieve. Regulations, which apply to all companies, really provide a level playing field for everyone by incorporating environmental costs into the price of doing business. In economic terms, such regulation internalizes the costs of environmental protection (making them part of the price a consumer pays for the product) as opposed to externalizing the costs (making all citizens pay the costs later in the form of higher taxes for environmental cleanup or public health services). The playing field is only completely level when everyone has to play by the same

A postcard from around 1910 from the oil region of western Pennsylvania (encompassing the towns of Franklin, Oil City, Titusville, Warren, and others). This card is typical of many from that time period, reflecting pride in the accomplishments of a dynamic young industry. Individual vignettes on the card show oil erupting from a torpedoed well, a tank fire, and somewhat more environmentally benign views of derricks and refineries.

rules, and because many environmental regulations are formulated on a state by state basis, that level field does not exist for the nation as a whole, let alone the world.

Nonetheless, regulation remains important (and is a major focus of this conference), because it is not reasonable to expect that there will be less environmental disturbance associated with mining in the future than in the past. On the contrary, we have already mined the richest and easiest-to-mine deposits. Future minerals exploitation will necessarily involve lower-grade ores and deeper, or otherwise harder-to-mine deposits. In our copper mines, for example, we are now mining ore containing only 0.1–0.3 percent copper (and that represents the ore, not all the associated waste rock that must be removed to get to the ore). Mining lower-grade ore will entail blasting and moving more rock and creating more environmental disturbance. But offsetting that, we now know much more about how to mine in ways that protect the environment. Can we mine a deposit like Questa's molybdenum ore today in a sustainable fashion that meets all current and future environmental concerns? The answer is a resounding "maybe."

The answer is "maybe" because we clearly cannot anticipate every potential future consequence of present day actions (any more than the citizens of the early twentieth century were able to predict the environmental consequences of the industrial activity of their era). The maybe also comes because we may not have the will or the economic resources to finance projects that preclude all pollution. But we can take (and in many cases, have taken) sensible and cost effective steps in that direction. Requiring impermeable liners around and beneath potential sources of pollutants, clay caps above such sources to prevent water infiltration, and similar steps can prevent transport of virtually all potential pollutants, not just those elements for which we have current concerns.

Beyond environmental issues there remain the social questions surrounding the aesthetics of mines and mining. Because of the nature of ore-forming processes, metallic ore deposits generally occur in mountainous terrain, not in flatland areas. Thus, a conflict between the aesthetics of pristine mountain regions and mines will always exist. Mines and their infrastructure of haul roads, waste rock piles, stockpiles, mills, and tailings ponds are no longer seen by many, perhaps most, citizens as acceptable consequences of mining. Papers submitted for this volume from environmental groups go so far as to say that no responsible mining company should even submit plans for mines in the vicinity of parks, monuments, or wilderness areas. We are, however, willing to accept sprawling clusters of fast-food restaurants, motels, condos, and housing developments at or near the entrances to many of our parks. Denver, Salt Lake City, Colorado Springs, Aspen, Las Vegas, Albuquerque, and other large cities in the West spread their smog throughout the region and consume vast amounts of habitat in proximity to gorgeous mountains, and we accept and

A postcard from around 1900 showing a forest of oil derricks in an otherwise deforested landscape near Titusville in western Pennsylvania. The grim vista, from an area close to the original Drake well discovery in 1859, is captioned as a "scene of early oil excitement."

even encourage that growth. Why are sprawling cities and towns and their pollution acceptable, whereas mines and their attendant impact on scenery unacceptable?

I am not advocating despoiling the surroundings of parks. I am questioning the values that allow urban sprawl but disallow mining of major mineral deposits. People can settle elsewhere, but minerals must be mined where the deposits are found.

At some point we will need to assess where we are going as a nation and how we plan to retain our economic viability. The industrial zeal of previous centuries is clearly gone. But how will we maintain the standard of living that we all love while producing few raw materials and even fewer manufactured goods? We cannot live on imports and service industries alone, and current trends of declining production and soaring international trade deficits are clearly unsustainable in the long run.

As we work toward creating a nation (and a state) that has safe pollution standards and that maintains its scenic beauty and natural habitat, we also need to develop a better concept of where mining and manufacturing fit into that picture. Our regulations need to protect and conserve natural settings, but they also need to recognize the unique opportunities provided by world-class mineral deposits such as those at Questa and Silver City. Even more importantly, we need to find ways to consume fewer resources, recycle and reuse materials effectively, and minimize the frequent "conserve versus develop" clashes that result from rampant consumption. European nations are far ahead of us in demanding recycling of consumer products and reuse of raw materials, a natural consequence of having a far higher population density than the U.S. More scientific information and the development of new technologies will clearly be needed to guide us in finding the balance between these three end members: conservation of our habitats, development of our natural resources, and sustainable reuse and recycling of industrial materials.

CHAPTER ONE

THE PHYSICAL AND HISTORICAL FRAMEWORK

DECISION-MAKERS
FIELD CONFERENCE 2005
Taos Region

CHAPTER ONE

Rio Grande gorge and Taos Mountains.

DECISION-MAKERS FIELD GUIDE 2005

The Geology and Landscape of the Taos Region

Paul Bauer, *New Mexico Bureau of Geology and Mineral Resources*

With the geologist lies the special responsibility and opportunity of revealing the earth in all its beauty and power. ... If geology and the geologist neglect interpretation of the earth to society, they are guilty of relinquishing what should be one of their major contributions. —David Leveson, *A Sense of the Earth*

Taos is located at the edge of radically different landscapes. The town is situated on the eastern edge of a broad, high valley known as the Taos Plateau. The Taos Plateau forms the southern part of the San Luis Basin, part of the Rio Grande rift. The basin is flanked on the east, south, and west by highlands of the southern Rocky Mountains. To the east is the Taos Range of the southern Sangre de Cristo Mountains. To the south are the Picuris Mountains, a westward prong of the Sangre de Cristo Mountains. The western highland is called the Tusas Mountains or the Brazos uplift. Geologic processes, working steadily through many millions of years, have contributed to the formation of these contrasting landscapes.

The quietude of the Taos Plateau belies the tumultuous geologic history of the region. Taos is located in one of the most dynamic and stimulating geologic settings on the planet. Buried beneath the plateau is an enormous fissure in Earth's crust (the Rio Grande rift) that is six times deeper than the Grand Canyon and thirty times larger than the Rio Grande gorge. The rift continues to be active, as indicated by the common occurrence of small to moderate earthquakes and the presence of large chambers of molten rock at depth. Evidence of past catastrophic seismic and igneous events is visible from most anywhere in the Taos area: young fault scarps where the mountains join the plateau, and a profusion of volcanoes and lava flows in the basin. Fortunately for us, neither destructive earthquakes nor volcanic eruptions have occurred in the historic past, although both will undoubtedly occur in the future.

The mountains, valleys, and volcanoes convey a tale of past landscapes that have evolved dramatically and repeatedly during the last 2 billion years, from shallow tropical seas to vast, sand-duned deserts; from enormous white sand beaches to muddy meandering rivers and lush fern forests; and from enormous mountain ranges to flat, featureless plains. These geographic features—the bold mountain escarpments, the fertile, flood-prone valleys, the deep gorge, and the dry tablelands—have greatly influenced 10,000 years of human occupation.

Landforms and Landscapes

A landform is any natural, discrete surface feature with a characteristic shape. A landscape, however, is a unique, distinct cluster of landforms. Mountain slopes, mountain valleys, and mountain peaks are landforms, whereas mountain belts are landscapes; volcanoes, calderas, craters, and lava plains are landforms, whereas volcanic fields and volcanic plateaus are landscapes.

All landforms have geological underpinnings. Most landforms are formed by a sequence of highly complex geological systems or processes, many of which are not visible or evident at the surface of the earth. Furthermore, all landforms have been modified to some extent. Some landforms in the Taos area, such as the 2.7-million-year-old Ute Mountain volcano, have maintained most of their original shapes. Other landforms have been significantly altered from their original forms by the processes of weathering and erosion. The Taos Plateau has been extensively modified by the Rio Grande and its tributary rivers to form the modern canyon system.

It is important to make a distinction between the age of a landform and the age of the materials that compose the landform. The rocks that form Wheeler Peak are nearly 1.8 billion years old, whereas the uplift of the Sangre de Cristo Mountains is about 60 million years old, and the dramatic crags of the Wheeler Peak area were sculpted by glaciers that melted only about 12,000 years ago.

GEOLOGY OF THE MOUNTAIN RANGES

The three mountain ranges that surround the Taos Plateau are the southernmost surface expression of the Rocky Mountains. In New Mexico, this area has a long history of repeated uplift, erosion, and burial. The latest uplift began approximately 70 million years ago during the Laramide orogeny and continues today. In the Taos area the Sangre de Cristo Mountains expose three principal ages of rocks:

- Precambrian metamorphic rocks (1.8–1.4 billion years old)
- Pennsylvanian sedimentary rocks (about 300 million years old)
- Tertiary igneous rocks (26–18 million years old)

The Tusas Mountains contain only the Precambrian and Tertiary assemblages, whereas the Picuris Mountains are composed entirely of Precambrian rocks.

In general, Precambrian rocks are dense and strong and highly resistant to erosion. Consequently, in the Taos area, most of the highest mountain peaks and steepest slopes are composed of such rocks. They include a wide variety of rock types (granite, gneiss, quartzite, schist) and ages, including the oldest dated rock in New Mexico, near Wheeler Peak.

Even though Precambrian rocks represent more than 80 percent of the history of our planet since its formation 4.5 billion years ago, our knowledge of Precambrian geology is woefully incomplete. Precambrian rocks now exposed at the surface in the Taos area were deeply buried during much of Precambrian time, and we know little of what was happening at the surface. We do know that the crust was subjected to intense metamorphism, folding, faulting, and melting during the lengthy formation of the North American continent until about 1 billion years ago. These ancient rocks are easily seen in many places near Taos: at Taos ski valley, Ponce de Leon hot springs, Red River Pass, on top of Wheeler Peak, in the steep cliffs near Pilar, and in the hills behind the Ojo Caliente hot springs.

Precambrian rocks supply much of the mineral wealth in the Southwest, and such is the case in the Taos region. The distinctive, micaceous clay prized by Picuris Pueblo potters is derived from a Precambrian schist from the Picuris Mountains. The controversial mica mine south of Taos is located in a similar deposit. Precambrian quartzite hosts the rich copper veins at Copper Hill near Peñasco. Gold was mined from Precambrian rocks near Red River. Copper was mined at Twining (now Taos ski valley) around 1900. The schists of the Picuris Mountains contain the "fairy cross" staurolites sold in local mineral shops. The Harding pegmatite mine near Dixon is a world-famous mineral collecting locality and has produced important amounts of the rare metals lithium, beryllium,

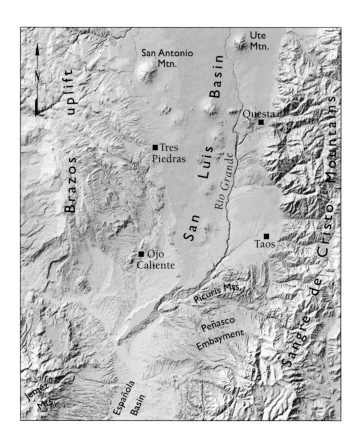

Physiographic map of the Taos region.

tantalum, and niobium. Suggestions of gold and silver deposits have periodically piqued the interest of prospectors and speculators in the Hopewell Lake area of the Tusas Mountains.

From 1.4 billion years ago until the Mississippian Period (354 million years ago) there is no rock record to study. Where rocks are absent, geologists must determine whether sediments were not deposited, or whether sediments were deposited and later eroded. In general, an advance of the sea promotes sedimentary buildup, whereas a retreat of the sea and exposure of the landmass to the air promotes erosion. Because sediments of Cambrian to Devonian age are absent in northern New Mexico but present to the south, geologists infer that during that time the Taos area was a landmass, and southern New Mexico was ocean.

During Pennsylvanian time, 80 percent of New Mexico was submerged beneath warm, shallow, equatorial seas. Northern New Mexico probably resembled modern-day North Carolina, with a flat coastal plain, swampy forests, and high mountains to the west. Mud, silt, sand, and gravel were eroded from nearby island highlands into a large sedimentary basin known as the Taos trough. Over time, as the thousands of feet of sediments were buried and compacted, they evolved into the sedimentary rocks that we now see east and south of Taos. These limestones, siltstones, shales, sandstones, and conglomerates are exposed over most of the southern part of the Sangre de Cristo Mountains, with excellent exposures on NM–64 along the Rio Fernando and along NM–518 south of Talpa.

A close examination of the Pennsylvanian rocks exposed near Taos reveals evidence of their original environments of deposition (rivers, deltas, shorelines, tidal flats, shallow seas), including features such as ripple marks, raindrop imprints, and crossbeds. The warm Paleozoic seas teemed with ancient life, and paleontologists have classified the fossilized remains of hundreds of marine species in the Taos area, including many varieties of clams, snails, sea lilies, corals, and brachiopods.

During Pennsylvanian time the region was subjected to a major mountain-building event known as the Ancestral Rocky Mountain orogeny. Today the most apparent effects of this are tilted and faulted Pennsylvanian strata. The modern Rocky Mountains mimic the chain of mountains that developed in Pennsylvanian time, because once the crust is broken it remains a zone of weakness to be exploited by later pulses of mountain-building events.

Mesozoic rocks do not exist in the Taos area, although they are exposed in the Moreno Valley near Angel Fire. The Triassic and Jurassic Periods (248–144 million years ago) were characterized by deposition of nonmarine sandstone and shale over much of northern New Mexico. Until very late in Mesozoic time, this area was alternately land and sea, as the shoreline advanced and retreated over a low-relief landscape. Great thicknesses of shale, sandstone, and limestone accumulated in these vast Cretaceous seas, and at times, much of the state was submerged. Sharks teeth are common Cretaceous fossils in New Mexico. Dinosaurs roamed the lush landscape of shorelines, river valleys, and widespread swamps for over 100 million years, until the great extinction 65 million years ago.

From Late Cretaceous time to early Tertiary time the Taos region was squeezed by forces associated with the next great mountain-building event, the Laramide orogeny. Compression of the crust resulted in major uplift and erosion and development of folds and faults. The western half of the Taos trough was uplifted and eroded, and by Tertiary time only rolling hills remained. Later, during Oligocene time, the San Luis Basin began to subside, while the Sangre de Cristo Mountains began to rise. Volcanoes erupted in the uplifted area, covering the region with lava and ash. The largest volcanic event was the explosive eruption of the Amalia Tuff from the Questa caldera about 25 million years ago, approximately coincident with the onset of the next great event, Rio Grande rifting.

The Bear Canyon pluton, visible in Red River Canyon, is one of many magma bodies that were emplaced in the Questa area following the caldera eruption. Collectively, these igneous rocks make up the "Questa magmatic system." As the magmas cooled and solidified, mineral-rich hydrothermal fluids percolated through the rock, and minerals such as molybdenite, quartz, and pyrite were precipitated in small fractures. These fracture-fillings (veins) later became the targets of prospectors and miners. Veins of the mineral molybdenite (molybdenum sulfide), an ore of the metal molybdenum, are found in unusually high concentrations in the Questa/Red River area and are now mined by Molycorp for use as lubricants and in the manufacture of stainless steel.

During the past 20 million years, the mountains have continued to rise and erode as the continental crust has continued to extend. By Pleistocene time (1.8–0.01 million years ago), the Sangre de Cristo Mountains had mostly developed their modern form.

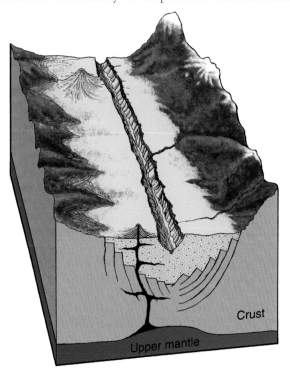

Block diagram showing the principal geologic elements of the Rio Grande rift in the the Taos area. A deep sedimentary basin (San Luis Basin) has formed in the crust above a bulge of hot asthenosphere. The Rio Grande has eroded a deep canyon in the rift basin. Not to scale.

CHAPTER ONE

View northeast of the two principal elements of the Taos landscape: the Sangre de Cristo Mountains and the Taos Plateau and gorge. The gorge here contains Pleistocene landslides of the Pliocene Servilleta Basalt.

During the Pleistocene ice ages, glaciers covered the higher peaks, scouring out depressions called cirques, and dumping huge volumes of sediment-laden water into the San Luis Basin.

GEOLOGY OF THE RIO GRANDE RIFT, SAN LUIS BASIN, AND TAOS PLATEAU

Taos is situated in one of the few young continental rift valleys on Earth. The other great rift valley is the 4,000-mile-long East African rift, which, as it has torn eastern Africa apart, has created Africa's highest peaks, deepest lakes, and equally impressive landscapes. The Rio Grande rift is part of a global system of fractures in Earth's uppermost rigid layer (the lithosphere) that have formed in order to accommodate relative movements of the lithospheric plates. The lithosphere is broken into about a dozen major plates that move over an underlying layer of partially molten rocks (the asthenosphere). Although most of the activity associated with plate tectonics occurs at or near plate margins, in some cases, such as along the Rio Grande rift, activity occurs well within the continental plate.

For the last 30 million years, plate tectonic forces have slowly begun to tear the North American continent apart along the Rio Grande rift. The lithosphere has been uplifted, stretched, thinned, broken, and intruded by magma. The resulting rift valley has sliced New Mexico and half of Colorado in two for a distance of over 600 miles. Near Taos the rift basin is about 20 miles wide and approximately 16,000 feet deep. The Rio Grande flows southward through successive rift basins that are linked by geologic constrictions. The river itself did not excavate the rift; the river follows the topographically lowest path along the rift, from the San Juan Mountains in Colorado to the Gulf of Mexico.

The transition from the basin to the surrounding mountains is abrupt, and in places such as near Taos and Pilar, is marked by steep slopes and cliffs. These abrupt transitions are both physiographic and geologic boundaries; most are major fault systems that delineate the margins of the rift. The eastern fault system is known as the Sangre de Cristo fault. The southern fault (the Embudo fault) separates the San Luis and Española rift basins.

The rift basin is filled with young (less than 30 million years old) materials, principally of sediments shed from the surrounding mountains, transported southward by the Rio Grande, and volcanic rocks of the Taos Plateau. We can see only the young basin fill at the surface, although material as old as 5 million years is exposed in the bottom of the Rio Grande gorge. The gorge is a much smaller and younger feature than the rift, and it is entirely erosional. The river began cutting down into the plateau sometime after 2.8 million years ago, which makes it younger than the Grand Canyon in Arizona. The deepest section of the gorge, at 850 feet, is located west of Questa in the Wild Rivers Recreation Area.

Visible from the rim of the gorge is a series of near-horizontal layers of basalt, known as the Servilleta Basalt. Most of the basalt was erupted as long, thin flows from topographically low volcanoes near Tres Piedras. The rift contains hundreds of such volcanoes, many of which are relatively young and retain their original conical shape. Nearly all of the isolated rounded hills that are scattered across the Taos Plateau are

Ute Mountain volcano rises above the basalt-capped Taos Plateau in northern Taos County. The Rio Grande excavated the Rio Grande gorge during the last 600,000 years.

volcanoes that erupted from 6 to 2 million years ago.

The rift basin is surrounded by alluvial fans that have advanced from the mountains into the basin. An alluvial fan begins to form where a rapidly moving mountain stream flows out onto a relatively flat valley floor. As the stream suddenly loses velocity, the coarsest sedimentary material is dropped by the stream. This material forms an "apron" that radiates out from the point where the mountain stream enters the valley. Most of this clay, sand, and gravel, called the Santa Fe Group, was eroded from the mountains during the past 30 million years. Over time, as the basin subsides alluvial fans are buried under successively younger alluvium. The youngest of these alluvial fans sustain the many sand and gravel quarries in the Taos Valley.

Many thousands of feet of rock were removed from the mountains and transported into the valley as the mountains slowly pushed upward. Similarly, the Rio Grande has deposited mineral-rich sediment into the basin for millions of years. Much of the thick sedimentary material in the Santa Fe Group is porous and permeable, and therefore serves as the principal aquifer in the region. People, agriculture, and industry have tended to concentrate along the Rio Grande rift for its fertile floodplain soils and precious supply of water.

Today, as uplift of the mountains and subsidence of the basin continue, streams persist in moving weathered rock from the mountains to the basin. The uplift takes place episodically through a series of small movements, each one associated with an earthquake. Although these processes work so slowly that we see little if any change during our lifetimes, over geologic time the countless small changes translate into a never-ending cycle of magnificent landscapes that are successively reduced to sand grains and washed to the sea.

Hydrology and Water Supply in the Taos Region

John W. Shomaker, *John Shomaker & Associates, Inc.*
Peggy Johnson, *New Mexico Bureau of Geology and Mineral Resources*

The Sangre de Cristo Mountains occupy the eastern half of the Taos region. Runoff from the steep slopes of Precambrian-age igneous rocks and Paleozoic rocks, largely limestone, supplies significant quantities of high quality water to the Taos Valley, recharges the region's aquifers, and provides most of the water used in the region. This paper provides a brief summary of water supply and water use in the Taos region.

WATER SUPPLY

Water in streams and aquifers in the Taos region is important both locally and to downstream water users along the middle Rio Grande. In 2000 more than 92 percent of all water diversions in Taos County (which includes surface and ground water diverted for all uses) originated from surface water sources, and more than 97 percent of the region's domestic and public supplies originated from ground water. In short, surface water supports the region's agriculture and economy, whereas ground water provides almost all of the region's drinking water. The Taos region generates significant surface water resources through eleven perennial streams and rivers, which are important to local as well as downstream users, and stores large quantities of ground water distributed in various local and regional-scale aquifers, some of which have been developed. These two regional sources, surface water and ground water, are intricately linked, and both discharge to the region's principal hydrologic feature, the Rio Grande. The Rio Grande in Taos County is largely unavailable for use because its course is in a deep gorge through much of the region, and because most of its flow is committed downstream.

The entire Taos region is drained by the Rio Grande. The divide between the Rio Grande and Canadian River watersheds forms the county's eastern boundary. The Rio Grande brings an annual average of about 325,500 acre-feet across the state line from Colorado, but almost all of the water used in the region is supplied by the tributaries, which also contribute to the Rio Grande. The annual outflow from the region, as measured at the Embudo gage, averages 601,700 acre-feet. Thus, the Taos region contributes an average 276,200 acre-feet each year to the Rio Grande, which is more than two-thirds of New Mexico's maximum annual allocation of water under the Rio Grande compact.

As the perennial streams of the Taos region exit the high crystalline-bedrock valleys of the Sangre de Cristo Mountains and enter the lower alluvial valleys, a portion of their flow is generally lost to infiltration through the coarse sand and gravel of alluvial fans and slopes. Although this loss diminishes the stream's discharge, it in turn replenishes storage in the region's aquifers and sustains shallow water levels in the region's valleys. Mountain-front faults and low-permeability aquifer units can force shallow ground water to the surface in some localities, producing springs, seeps, and marshes and rejuvenating stream flow. The complex hydrogeologic conditions adjacent to the mountain front give rise to complicated interactions between water flow in streams and water flow in aquifers. As the region increases its reliance on ground water, withdrawals from the shallow aquifers via wells will eventually reduce stream flow by intercepting water that would otherwise maintain the streams or by drawing water directly from the stream channels.

Precipitation in the region is strongly influenced by elevation. Average annual precipitation ranges from less than 10 inches at Ojo Caliente (6,290 feet) to more than 20 inches at Red River (8,680 feet). Monthly precipitation is generally greatest in July, August, and September. Runoff in the Rio Grande tributaries, on the other hand, reaches its peak in April or May with melting of the accumulated winter snow pack. Both precipitation and stream flow conditions are highly variable from year to year, and "average" conditions are the exception rather than the rule. In general, historic stream flow fluctuates between periods of above-normal and below-normal discharge, separated by short periods of transition. The most severe dry conditions occurred between 1950 and 1964, when annual discharge for this fifteen-year period averaged only 341,000 acre-feet or 57 percent of normal. The wettest period on record occurred during the strong El Niño events of the mid-1980s, which

THE PHYSICAL AND HISTORICAL FRAMEWORK

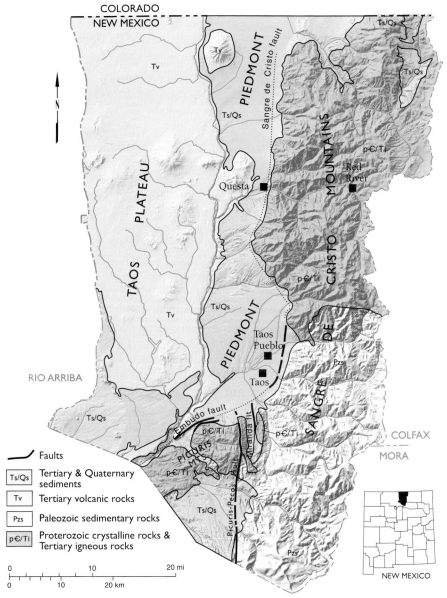

Generalized geologic map of Taos County.

produced a five-year average flow that was 180 percent of normal. (Given the precipitation records to date for this year, 2005 may become the wettest year on record.

WATER USE

Apart from small diversions and shallow wells in the narrow higher-elevation valleys, the region's water use takes place in the lower valleys adjacent to its perennial streams and along the east side of the Rio Grande gorge. About 99,557 acre-feet (97.9 percent of it surface water) was diverted for irrigation in the Taos region in year 2000, which in turn accounted for 92.7 percent of all the water withdrawn. Sixty-one percent (60,289 acre-feet) of the water diverted for irrigation was returned to streams. Water use for mining in year 2000 was 3,094 acre-feet, about 2,568 acre-feet of which (83 percent of the water diverted for mining) became return flow, available to other users. Overall, 60 percent of the water diverted for use in Taos County is returned to the streams and aquifers and remains available for other users.

Some 2,255 acre-feet, almost all from ground water, was withdrawn for public supply in 2000, and another 1,376 acre-feet was pumped from domestic wells. Ground water pumping for public and domestic supply roughly doubled in the fifteen years between 1985 and 2000. The largest user is the town of Taos, pumping roughly one-third of the public supply; the remainder is produced by the El Prado Water and Sanitation District and many small mutual-domestic systems. About 59 percent of the water pumped for public supply was returned to streams as treated wastewater; virtually all of the water from domestic wells is assumed to have been lost to evaporation.

WATERSHEDS

The perennial streams and rivers within the Taos region occupy large to intermediate drainage basins on the western face of the Sangre de Cristo Mountains. The principal watersheds in the region are summarized here:

Taos Valley

The Taos Valley includes two major drainages, Arroyo Hondo and the Rio Pueblo de Taos. The Rio Pueblo is sustained by five major tributaries: Arroyo Seco, Rio Lucero, Rio Fernando de Taos, Rio Grande del Rancho, and Rio Chiquito. The combined drainage area of these streams is about 530 square miles. Much of the discharge in these streams is diverted for irrigation in summer, but a combination of return flow from irrigation and winter flows provides recharge to a complex aquifer system and augments flow in the Rio Grande.

The water table is at land surface in much of the eastern part of the valley, creating Taos Pueblo's Buffalo Pasture and other marsh areas and supporting innumerable small seeps and springs. These occur where the highly permeable, shallow aquifer thins, loses storage capacity, and is unable to transmit the full amount of the recharge entering the aquifer along the foot of the eastern mountains. The water table has declined in recent years, partly in response to drought and presumably also due to increased ground water pumping. The surface area of the Buffalo Pasture is said to have decreased, some springs have stopped flowing, and many shallow wells have failed.

Recharge moves westward away from the mountain front, and downward through a sequence of basalt flows and interbedded sediments close to the surface, then through sandstone, mudstone, and conglomerate beds of the Tertiary-age Santa Fe Group aquifer before discharging into the Rio Grande. The Rio Grande gorge at the western edge of the Taos Valley is 500–600 feet deep, and the 700–900-foot elevation difference between the areas of recharge and the Rio Grande, over the relatively short distance of 7–9 miles between the river and the mountain front, leads to a strong downward component of ground water flow, and probably to perched ground water and unsaturated conditions below the water table in many places. Recharge to the Taos Valley aquifer system probably averages about 37,500 acre-feet per year, a third of which is in the form of seepage from irrigation. About 15,000 acre-feet per year, or 21 cubic feet per second, discharges to the Rio Grande through seeps and springs, and flow through the channel bottom, between the mouth of Arroyo Hondo and the Taos Junction gage.

Pumping tests of wells indicate ranges in flow from a few tenths of a foot per day to 22 feet per day for the alluvium, from about 5 feet to more than 25 feet per day for the underlying Tertiary-age basalt and associated sediments, and from a few tenths of a foot to about 2 feet per day for the still deeper Tertiary-age Tesuque Formation beds in the basin fill. Well-yields vary from place to place, but pumping rates as high as

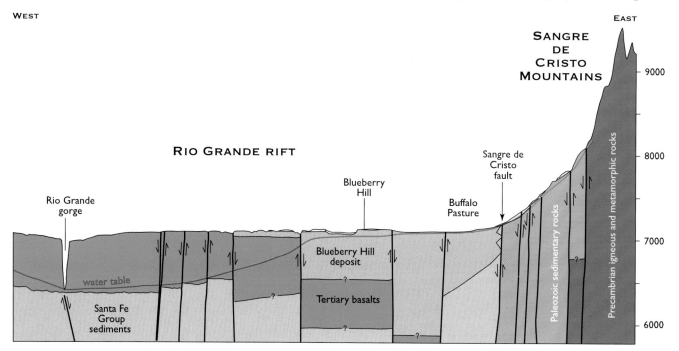

Geologic cross section showing approximate location of the water table in the Taos region. Recharge from the mountains infiltrates and moves westward toward the Rio Grande. Ground water flow is influenced by the geologic materials in the subsurface and by geologic structures, such as faults. Colors correspond to the geologic map on the inside back cover. Note that the vertical scale is exaggerated in order to show topography. Courtesy of Paul Bauer, from unpublished data.

several hundred gallons per minute have been reported for wells tapping the relatively shallow alluvium. Rates for wells in the underlying basalt and associated sediments have been as high as 120 gallons per minute and, for the still deeper beds, as high as 500 gallons per minute.

Pumping from wells affects the flows in the streams near the mountain front. Because of the strong downward flow and potential for some disconnect between shallow and deep ground water, it is likely that future wells will be deep, cased through the upper part of the aquifer, and preferentially located near the Rio Grande, in order to minimize both the effects on the streams and drawdown in the many shallow wells. Wells are commonly less than 200 feet deep in the eastern part of the valley, but must be more than 800 feet deep near the Rio Grande to penetrate below the water table. There are an estimated 1,900 private domestic wells in the Taos Valley.

The combined average yield of the Taos Valley streams is estimated at about 94,000 acre-feet per year, varying from about 80 percent of that (or 76,000 acre-feet per year) in "dry" years to 138 percent of that (or 129,000 acre-feet per year) in "wet" years. (Dry and wet years are defined by the 25th and 75th percentile, respectively, of flow in Arroyo Hondo and the Rio Pueblo de Taos.) Ditches on Taos Pueblo lands plus some fifty-five acequias on non-pueblo lands irrigate about 14,000 acres along the eastern side of the valley. Although the average annual flows of the streams that serve these ditches and acequias are theoretically sufficient to supply them, the variability from year to year, and the concentration of flows during the spring and early summer, lead to shortages during late irrigation season in most years. There is no reservoir storage on most streams.

Sunshine Valley

The Sunshine Valley is a high plain lying between the Sangre de Cristo Mountains and the Rio Grande, and between Questa and the Colorado state line. Shallow ground water pumped from alluvium, Santa Fe Group basin fill, and interbedded basalts irrigates about 600 acres (year 2000). There are no perennial streams. Recharge to the Sunshine Valley aquifer has been estimated at 20,000 acre-feet per year. If there were no ground water pumping, this amount would discharge from the aquifer system to the Rio Grande. Well yields in the Sunshine Valley are generally as high as 1,200 gallons per minute, but a 3,000-gallons-per-minute well has been reported. The high permeability of the alluvium and basalts, which accounts for the high yields, also leads to large depletion of flow in the Rio Grande as a result of pumping.

Costilla Creek Valley

Costilla Creek drains the northern Sangre de Cristo Mountains, yielding an estimated average of about 29,000 acre-feet per year. Costilla Reservoir, at almost 9,500 feet in elevation and with a nominal capacity of 16,500 acre-feet, regulates deliveries. Eighteen acequias serve about 5,500 acres of irrigation, and also deliver water to irrigators in Colorado and the Sunshine Valley. Ground water use is limited to domestic and livestock supplies, and about 100 acres of irrigation.

Red River Valley

The Red River watershed yields an estimated average of about 50,400 acre-feet per year. The thin, narrow body of alluvium in the bottom of the valley and fractured bedrock close to the river are sufficiently permeable to support relatively high yield wells, but their production is, in effect, diversion from the river. About 3,100 acres are under irrigation from the Red River and its tributary Cabresto Creek. The Molycorp operations above Questa represent the only other large diversion. The town of Red River diverted about 87 acre-feet of surface water in 2000 and pumped about 487 acre-feet of ground water for public supply.

This summary is based on published reports, but water rights in the Taos Valley are in the process of adjudication, and much new work, confidential as of this writing, has been done to refine the inventory of irrigated acreage, the understanding of the flows in streams and the ground water system, and the expected pattern of future water use.

Suggested Reading

Drakos, P., Lazarus, J., White, B., Banet, C., Hodgins, M., Riesterer, J, and Sandoval, J., 2004, Hydrologic characteristics of basin-fill aquifers in the southern San Luis Basin, New Mexico; in Brister, B. S., Bauer, P. W., Read, A. S., and Lueth, V. W. (eds), Geology of the Taos region: New Mexico Geological Society, Guidebook 55.

Garrabrant, L. A., 1993, Water resources of Taos County, New Mexico: U.S. Geological Survey, Water Resources Investigations Report 93-4107.

A History of Agricultural Water Development in the Taos Valley

Peggy Johnson, New Mexico Bureau of Geology and Mineral Resources

The agricultural tradition in the Taos Valley began hundreds of years ago with Pueblo culture, expanded with Spanish settlement, and continues today as an important economic foundation for rural communities in the valley. Irrigated agriculture is sustained in Taos, and throughout much of the Rio Grande valley, by stream flow originating in the Sangre de Cristo and Picuris Mountains east and south of the Taos Valley. The ancient and historic settlement patterns that created the present-day population centers surrounding Taos Pueblo, Taos, El Prado, Ranchos de Taos, Ranchitos, Cañón, and Arroyo Hondo among others, were determined by the abundance of fertile land and irrigation waters surrounding the valley's major streams.

Although irrigation in the Taos Valley is typically associated with the Spanish acequia culture, the practice of irrigated agriculture was initiated by Pueblo farmers long before Spanish settlement. When the Spanish first arrived in the Taos Valley in 1540, they were notably impressed by the stature of Taos Pueblo. Narratives of the Coronado expeditions noted that all pueblos, including Taos, grew maize, beans, and squash, and observed that the Rio del Pueblo (Rio Pueblo de Taos) that divided the village flowed swift and deep. When Spanish settlers returned to the Rio Grande in 1598, Juan de Oñate admiringly described Pueblo agriculture to the Spanish king and viceroy, noting that the Pueblo Indians were skilled farmers. Some fields were irrigated, and others relied on summer rains. The region later proved particularly well adapted for irrigated wheat production, and bumper crops grown with a dependable water supply made the region New Mexico's breadbasket when shortages threatened elsewhere. Today 92 percent of Taos County's water requirements (107,342 acre-feet) are supplied by surface water, most of which is used for irrigated agriculture.

The issues of population growth, water demand, and over-allocation that dominate today's headlines and planning agendas first emerged with Spanish colonization on land surrounding Taos Pueblo. During the turbulent years before the Pueblo Revolt of 1680, early Spanish occupants settled along the rivers north and west of the pueblo to take advantage of the fertile

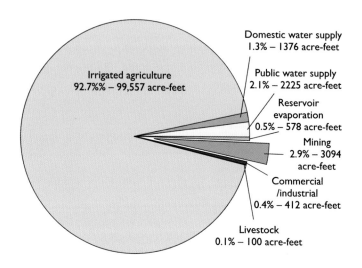

Distribution of water use in Taos County today.

lands and abundant water for irrigation. Post-revolt resettlement in the Taos Valley renewed requests for agricultural and grazing lands situated along the valley's major streams, resulting in land grants throughout the valley by the Spanish government. One of the earliest references to a specific acequia in the Taos area (September 1715) cites the Acequia de los Lovatos, the eastern boundary of the Gijosa Grant, which originates in the Rio Pueblo and is still in use today. Modern records document approximately eighty acequias and ditches operating in the Taos Valley, most of which date from Spanish development in the eighteenth and nineteenth centuries.

Early settlers and provincial officials knew that no community could exist without an assured water supply. As early as 1795, accelerating population growth caused increased competition between Hispanos and Pueblo Indians for water, and the early New Mexican water administrators (*cabildos or ayuntamientos*) developed various methods to adjust land policy with water availability in a semiarid environment. Hence began the practice of water administration that still embroils the region today.

From the earliest Spanish chronicles of water resources to the recent state engineer assessments in the late twentieth century, observers have noted, quantified, and apportioned the prodigious but highly

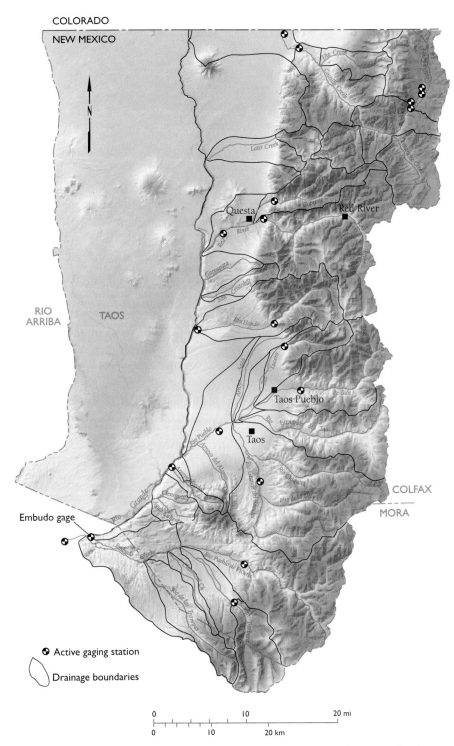

Drainage basins and streams in Taos County.

for irrigation and domestic use for the community of Questa. These six streams and their watersheds provide much of the water that sustains the Rio Grande on its course through New Mexico.

RIO PUEBLO DE TAOS AND RIO FERNANDO

The Rio Pueblo drains an area of 67 square miles in the Sangre de Cristo Mountains and produces an average annual discharge of approximately 22,000 acre-feet. Waters from the Rio Pueblo have been used by residents of Taos Pueblo for untold years before Spanish settlement, but the earliest written descriptions of the valley were made by Franciscan friars during visits to New Mexican missions. These early reports always noted the "fair-sized rivers" and the fields of wheat and corn cultivated by Taos Indians and irrigated from the Rios Lucero and Pueblo. The earliest Spanish settlement along the Rio Pueblo began downstream of Taos Pueblo with authorization of the Gijosa Grant on September 20, 1715. A population explosion during the 1790s caused increased competition for land and water between settlers and the pueblo and among the settlers themselves. A census in 1796 confirmed a Pueblo Indian population of 510 at Taos and 199 at Picuris, approximately one-half of the area's total population. The Hispano population had grown from 330 to 779 since the last census in 1790, a 135 percent increase in six years. The census indicated about ten families at each of the villages within the Gijosa Grant along the Rio Pueblo now known as Upper and Lower Ranchitos.

Despite being the smallest of the principal tributaries feeding the Rio Pueblo de Taos, the Rio Fernando irrigates many acres of crops and pasture. Dependent on springs and snowmelt, the Rio Fernando drains 70

variable flows in the major streams of the Taos Valley, which include the Rio Hondo, Rio Lucero, Rio Pueblo de Taos, Rio Fernando, and Rio Grande del Rancho. The Red River, which occupies a major watershed north of Arroyo Hondo, drains the area developed by the Molycorp molybdenum mine and provides water square miles below Palo Flechado Pass in the Sangre de Cristo Mountains, with a mean annual discharge of 4,140 acre-feet. Authorized Hispano settlement began in 1796 with the Don Fernando de Taos Grant, which encompassed lands adjacent to Taos Pueblo and bisected by the Rio Fernando. With sixty families taking residence in the first year, followed by a number of new arrivals a year later, the Fernando Grant quickly became the largest Hispano community in the valley. In the fall of 1797, after only two planting seasons, a request was made by the new residents for rights to the surplus waters (*sobrantes*) from the Rio Pueblo and Rio Lucero, suggesting that the Rio Fernando had already proved inadequate for the settlers' needs. A *sobrante* right meant that the newcomers at Don Fernando could use any water remaining after Pueblo farmers had satisfied their requirements. For many years Hispano farmers managed to maintain good relations with the pueblo, and irrigators found ways to share the stream's water without resorting to lawsuits. In 1871, however, litigation changed the settlers' sobrante right into an absolute share, thus initiating an era of repeated controversies over apportionment between the pueblo and downstream users. The Abeyta adjudication represents the current apportionment suit.

An unusual water management case, possibly the first with environmental implications, occurred on the Rio Fernando in the spring of 1877. Several landowners who depended on the Rio Fernando for irrigation water complained that a certain Juan Sánchez had been systematically cutting down large numbers of cottonwood trees near the river's headwaters, thus eliminating the cooling shade that protected the stream. Exposure to the blazing sun, they claimed, would cause water shortages from evaporation. The presiding judge ruled in favor of the plaintiffs, finding that excessive timber cutting in the bosque would lead, little by little, to diminution of the water necessary for agriculture in the valley. Thereafter, anyone convicted of such destruction would be regarded as a transgressor, subject to all the rigors of the law. Although nineteenth century New Mexicans appeared generally unaware of environmental issues, this case demonstrates an intuitive understanding of the basic concepts of a hydrologic cycle and water balance applied by present-day hydrologists.

RIO GRANDE DEL RANCHO

The Rio Grande del Rancho, with tributaries the Rio Chiquito, the Rito de la Olla, and Arroyo Miranda, drains the south side of the Fernando Mountains and the north side of the Picuris Mountains. This fairly large watershed (150 square miles) is mostly coincident with the Rio Grande del Rancho Land Grant authorized in 1795 and includes the towns of Talpa and Llano Quemado. Diversion of stream flow below Talpa through nineteen acequias and ditches currently provides irrigation for more than 3,380 acres. A stream gage located on the Rio Grande del Rancho near Talpa has operated since 1953, recording a mean annual discharge of 15,340 acre-feet.

The oldest Spanish grant in Taos Valley, authorized on June 15, 1715, to Captain Cristóbal de la Serna, allocated a tract of agricultural and grazing land west of the Rio de las Trampas (Rio Grande del Rancho) near present Ranchos de Taos. The Serna Grant was later sold to Diego "El Coyote" Romero, developed successfully, and passed to his heirs. One of the earliest conflicts concerning water and land allocation occurred along the Rio Grande del Rancho when Ranchos residents protested upstream settlement by outsiders on the proposed Rio Grande del Rancho Grant early in 1795. Additional irrigation above their fields, they said, would inevitably diminish their share of the river's flow and endanger the livelihood of present landowners (and added that decreased harvests meant a corresponding decline in tithes and first fruits necessary for maintenance of the church). In an effort to thwart upstream settlement, the protestors petitioned the governor, asking that they themselves receive possession of the grant lands. The request was approved on February 4, 1795, thus consolidating control of the river's headwaters into the hands of the Ranchos residents.

THE RIO LUCERO AND ARROYO SECO

Arroyo Seco and the Rio Lucero begin in adjoining canyons south of the Rio Hondo and run southwest from sources high in the Sangre de Cristo Mountains to meet the Rio Pueblo near Los Cordovas and Upper Ranchitos. Settlement between the Lucero and the Hondo began before the Pueblo Revolt of 1680 and continued through the eighteenth century in a bewildering series of overlapping grants, including Antonio Martínez in 1716, Pedro Vigil de Santillanes in 1742, and Antonio Martín in 1745. Actual colonization at Arroyo Seco and Desmontes was delayed until the early nineteenth century, when the land between Arroyo Seco and Arroyo Hondo came under the control of Mariano Sanches, an aggressive land developer, who began a vigorous campaign to colonize the brush-covered flats. In a plan involving exchange of

The Embudo Gage and Stream Flow Measurement in Taos County

Embudo gage today.

These geologists measuring flow on the Arkansas River in Colorado ca. 1890 are using procedures developed at Embudo.

Modern efforts to measure discharge from the region's streams rely on a series of stream gages or measurement stations installed and administered by the U.S. Geological Survey (USGS) and the New Mexico Office of the State Engineer and Interstate Stream Commission. The oldest such station in the United States, active since 1890, is located on the main stem of the Rio Grande at Embudo station. In March of 1888, responding to a sudden interest in western irrigation, Congress passed a joint resolution authorizing the Secretary of the Interior to examine potential irrigated lands, locate possible sites for water storage, and determine the capacity of various streams. Responsibility for the project fell to John Wesley Powell, then serving as director of the U.S. Geological Survey. Recognizing a shortage of personnel trained in water measurement, Powell ordered the establishment of a school for hydrographers on a western river. The Rio Grande near Embudo station was selected because of its location in an arid region on a major stream unlikely to freeze in winter. For five months between December 1888 and April 1889, eight recent graduates from prestigious eastern engineering schools worked under Frederick Haynes Newell to learn a technique of water measurement known as stream gaging. The class made regular measurements of water temperature, depth, and velocity in the Rio Grande at Embudo station from a makeshift raft, thus providing the first such measurements recorded in the United States.

Historically, thirty-five such stations have operated between the New Mexico–Colorado state line and Embudo, although only twenty-one are currently active. These gages measure stream discharge (in units of cubic feet per second or cfs) on an hourly basis. Real-time data reflecting current conditions are recorded at 15 to 60 minute intervals, stored on site, and then transmitted to USGS offices every four hours. Recording and transmission times may be more frequent during critical events. Data from real-time sites are relayed via satellite, telephone, or radio and are available for viewing within three minutes of arrival. Online data for Embudo station can be found at **http://nwis.waterdata.usgs.gov/nwis/nwisman/?site_no=08279500&agency_cd=USGS**.

What began at the Embudo gage in 1889 developed into a program encompassing approximately 1.5 million sites in all fifty states, the District of Columbia, and Puerto Rico. This network of stream gage stations provides a critical dataset that allows monitoring during critical flood events, supports a statistical assessment of the amount of water available for diversion and use, and facilitates water management and administrative decisions.

land for services, the entrepreneur offered potential farm sites to workmen (*trabajantes*) enlisted to clear (*desmontar*) the plain's thick growth of sage, piñón, and juniper. Named by the brush cutters, Desmontes grew up as a scattered settlement on the high, dry plains north and west of the Arroyo Seco. Not surprisingly, Arroyo Seco provided only a bare minimum of water for irrigation, causing the trabajantes to cooperate in the construction of two major canals that divert waters from other streams. Each one represents a formidable project; both are still in use. The Acequia Madre del Rio Lucero del Arroyo Seco diverts water from the Rio Lucero and runs it for more than a mile across the dry plain to Arroyo Seco Creek, where it follows the channel downstream for later diversion or crosses the creek for distribution through a network of smaller ditches totaling several miles in length. The main canal, La Acequia de la Cuchilla, represents an even greater task. Diverting water from the Rio Hondo above present Valdez, the Cuchilla climbs up the south wall of Hondo Canyon for more than two miles before reaching the plain at the top, where it divides into several laterals that water fields at Desmontes. The ditch, which seems to run uphill, is still regarded

The Rio Hondo in the Wheeler Peak area.

with awe by professional engineers. However, Arroyo Seco's claims to water from two other rivers caused recurrent disputes with Arroyo Hondo, Taos Pueblo, and Fernando de Taos.

RIO HONDO

From its source among the high peaks of the Sangre de Cristo Mountains, the Rio Hondo rushes westward past the villages of Valdez and Arroyo Hondo before joining the Rio Grande 14 miles northwest of Taos. With an average annual discharge of more than 25,000 acre-feet, the river irrigates 3,700 acres through fifteen acequias. Official colonization in the Hondo Valley began in 1815 with authorization of the Arroyo Hondo Grant. Settlers along the Rio Hondo lived in harmony with their southern neighbors on the plain until the founding of San Antonio (today's Valdez) in 1823. While certifying Arroyo Hondo Grant papers in 1823, Alcalde Juan Antonio Lovato added language to the grant records in order to resolve the water dispute with Arroyo Seco. Lovato made the right of Arroyo Hondo village superior to that of San Antonio, and put both ahead of Arroyo Seco and Desmontes. Desperate for irrigation water, residents of the southern plain persuaded Arroyo Hondo to allow them a substantial part of the flow from their stream, a practice that became a matter of local custom. After New Mexico became a territory, citizens turned to the newly established court system to resolve water disputes. On May 19, 1852, Arroyo Hondo again brought suit in probate court against the inhabitants of Desmontes to determine rights of the two communities to water from the Rio Hondo. The presiding judge ruled that Arroyo Hondo had first priority, but allowed Desmontes one-third of the river's flow, even in time of scarcity. According to local residents this apportionment is still observed by the mayordomos of the acequias involved, with Arroyo Hondo's two-thirds divided between its own users and those of Valdez and Cañoncito.

Suggested Reading

Historic notes regarding Pueblo and Spanish agriculture and water development and early New Mexican water administration are taken from the two works of Baxter listed below. Information on water availability in the region's major streams is from the two Johnson references listed.

Baxter, John O., 1990, Spanish irrigation in Taos Valley. New Mexico State Engineer Office, Santa Fe, NM, 126 pp.
Baxter, John O., 1997, Dividing New Mexico's waters, 1700-1912. University of New Mexico Press, Albuquerque, NM, 135 pp.
Johnson, Peggy, 1998, Surface-water assessment, Taos County, New Mexico. New Mexico Bureau of Geology & Mineral Resources Open-File Report 440.
Johnson, Peggy, 1999, Availability and variability of surface-water resources in Taos County, New Mexico - an assessment for regional planning. New Mexico Geology, vol. 21, no. 1, February 1999.

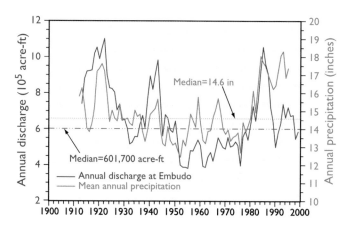

Five year moving average of annual discharge for the Rio Grande at Embudo, compared to mean annual precipitation for Cerro, Red River, and Taos.

Taos County's Mineral Heritage—A Unique Page in New Mexico's Mining History

Robert W. Eveleth, *New Mexico Bureau of Geology and Mineral Resources*

Mighty mountains, sparkling streams, valleys fair to look upon, set like precious stones in a region which combines the lovely and the lofty, the bold and the beautiful, the placid and the picturesque—this is the new El Dorado of Red River, Taos county, northern New Mexico.... –Anon 1897

The mineral riches and grandeur of Taos County that moved this unknown scribe to wax so eloquently have weighed heavily on the human imagination since prehistoric times, beginning in earnest soon after the arrival of the Spanish in 1540. However, various factors limited the development of its resources on a large scale well into the twentieth century. Remoteness and elevation were major adversities, then and now. The infrastructure usually associated with heavy industry, such as main and branch line railroads and factories, was totally lacking or long in coming. Moreover, the high country, one of the county's many endearing attractions, has always been difficult to access.

Native Americans took advantage of the mineral substances indigenous to the area for uses that they deemed necessary to their everyday lives: clay and mica for the manufacture of pottery, the various silica minerals for fashioning tools, knives, and weaponry, and certain colorful minerals for use in personal adornment and pigmentation. Historically speaking, gold placers, particularly at Arroyo Hondo, as well as other mines and prospects in the "Embuda" and Picuris areas, were said to have been worked by the Spanish long before the Pueblo Revolt in 1680. American miners often encountered evidence of their Spanish predecessors. Garnet "mines" were reported in the Picuris area around 1627 by Father Salmeron, and the ex-governor of Picuris Pueblo escaped the rather serious charge of witchcraft by disclosing the location of four veins of silver in Picuris Canyon in 1713. Prospectors during the Mexican period (1821–1848) worked the placers near what would become Amizette during the 1820s and 30s.

Taos County's many contributions to mineral discoveries and developments elsewhere in the Old

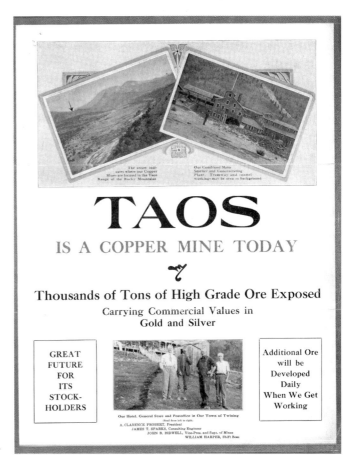

The Taos Mining Company (a successor to the Fraser Mountain Copper Company) was without doubt the most elaborate and costly attempt to develop the elusive metal deposits in the Red River district before 1920. The mill and smelter at the base of Frazer Mountain are pictured along with company officials Clarence Probert, James Sparks, John Bidwell, and William Harper. [Author's collection.]

West are also of significance. Many of the legendary pioneer trailblazers such as Kit Carson, Lucien Maxwell, and others were involved with mining and prospecting in Taos County and environs. Taos miner J. P. L. Leese arrived in California in 1833 or 34 with a considerable quantity of gold dust—gold that likely originated from the Arroyo Hondo placers, the small product of which had long been traded in Taos commerce and became part of the first gold shipment out of California.

CHAPTER ONE

The Denver & Rio Grande Western Railroad pushed its narrow-gage tracks south from Antonito along the western flanks of the Rio Grande through such garden spots as No Agua, Tres Piedras, and Embudo, thereby forever bypassing the more populous Taos. Thus Tres Piedras (shown here in this ca. 1910 view) and Embudo were Taos county's sole railroad shipping points. Had the railroad survived beyond 1941, it doubtless would have thrived on the No Agua perlite traffic. The long abandoned Tres Piedras water tank still stands today.

The instrument that opened the floodgates to the vast American West, including New Mexico west of the Rio Grande, was the 1848 Treaty of Guadalupe Hidalgo, which transferred ownership of much of Arizona and New Mexico from Mexico to the U.S. The new American proprietors immediately assumed the responsibility of dealing with the Apaches, Comanches, and others; they spent the next half-century engaging in military action to pacify the region. The American military provided some of the earliest accounts of rich mines and mineral resources. The early American press obligingly waxed eloquently on the existence of rich mines in the Taos area and elsewhere in the new territory and fired the imaginations of American prospectors. But the greatest influx of prospectors came with the close of the American Civil War.

MT. BALDY TO THE TAOS RANGE

Fort Union, near present day Watrous in Mora County, was the principal supply depot for the military infrastructure in New Mexico Territory and was of paramount importance in protecting travelers and the pioneer merchants along the Santa Fe Trail. All representatives of Old West society ventured to this outpost, including Native Americans who regularly traded with the prospectors stationed there. In 1866 a Ute stopped by with some high-grade copper specimens to barter for supplies and was directed to William Kroenig and others of the prospecting fraternity. Recognizing the potential value, the group either traded or paid the Indian for the specimens and, as part of the deal, were escorted to the locality near the top of Mount Baldy (near Elizabethtown) where they located the famous "Mystic Copper Mine." Additional prospecting soon revealed the presence of placer gold on Ponil Creek, and the rush was on. Elizabethtown was soon established; from there the miners fanned out to the west toward Taos and scrutinized the major drainages and peaks in search of more treasure. The few prospectors in the Taos area were doing much the same in reverse by tracing the placers eastward to their sources. Many promising discoveries followed, and miners were soon working the ground at Arroyo Hondo and Red River. During the next two decades localities such as Black Copper, Keystone, Anchor, Midnight, LaBelle, Gold Hill, Twining, and Amizette experienced the touch of the pick, shovel, pan, and sluice.

The arrival of the railroads in the 1880s brought immediate attention and development to New Mexico's vast coal resources: Magdalena's lead, silver,

The Independence mine on Bitter Creek, Red River, ca. 1905.

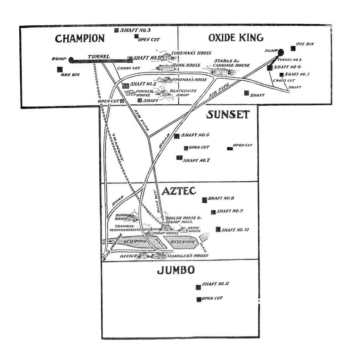

Sketch map of surface improvements at the Champion Copper Company property on Copper Hill, Picuris district, from the 1917 annual report. Most of the improvements should have been postponed pending the development of economically minable ore.

high-grade precious metal-bearing ores were occasionally encountered, which only further served to encourage the miners to pursue the inconsistent and elusive deposits so typically hosted by the Precambrian and Tertiary rocks in the Taos County area. Notably rich but woefully small pockets of gold ore were discovered in such mines as the Memphis and Independence (both on Bitter Creek). These discoveries too often resulted in premature, ill-advised investments in elaborate milling and, at least in one case, smelting facilities at properties such as the Caribel Group on Pioneer Creek, the Buffalo–New Mexico property on Placer Creek, the Champion Copper Company on Copper Hill, the Fraser (sic) Mountain Copper Company mine at Twining, and many others.

Arthur Montgomery and crew sorting beryl ore at the Harding pegmatite mine, Taos County, New Mexico, ca. 1952. The Harding was at times the nation's leading producer of beryl and the rare-earth minerals columbite-tantalite and microlite.

and zinc deposits, and southwestern New Mexico's base and precious metal districts. For a time Taos County was left far behind in the developmental scheme of things. The Denver & Rio Grande Western Railroad constructed its right of way down the west bank of the Rio Grande (thereby avoiding a very costly bridge across the gorge but forever bypassing Taos) through such thriving burgs as No Agua, Tres Piedras, and Embudo, on its way to a terminus in the Española Valley, the only real population center along the line. The inevitable talk of a branch line railroad was revisited as each little mining camp experienced its fleeting moment in the sun. The railroads wisely assumed a "wait and see" stance, with the result that Tres Piedras, Taos Junction, and Embudo were, through September 1941, the only real railroad shipping points Taos County ever had.

LATE NINETEENTH / EARLY TWENTIETH CENTURY MINING VENTURES

Despite nearly a century of prospecting, exploration, and sometimes extraordinary developmental effort, few if any mines and prospects in Taos County could be considered successful before 1920. Small zones of

Even with the extravagant efforts of the Fraser Mountain Copper Company (and its successor, the Taos Mining Company) and others, the mineral production record for Taos County before 1923 was approximately $100,000—a mere trifle of the amount invested. The small production of gold dust won by the pioneer miners went largely unrecorded, but regardless of all the hype, hope, and eternal optimism expressed by such camps as Amizette, LaBelle, Black Copper, and others, officially credited production is at or near zero. Events in the twentieth century would quickly change all that, and Taos County would finally enter the ranks of world-class mineral producers.

ALL THAT GLITTERS IS NOT GOLD (SOME OF IT IS MICA)

Taos County is unique among New Mexico's mineral producers for commodities such as beryl, columbite-tantalite, microlite, and one of the world's largest deposits of optical calcite at the Harding mine; mica (sericite) at the U.S. Hill (Tojo) mine; and molybdenum at Questa. Add to that the No Agua perlite, and the value of Taos County's mineral production *since* 1923 (exclusive of sand, gravel, and crushed stone products) approaches a half *billion* dollars.

The Harding pegmatite, located as early as 1900, went largely unappreciated until a miner from the Black Hills recognized in 1919 the true identity of the mineralogical suite. The lithium-bearing minerals were found to be of value in the manufacture of glass (and Edison's alkaline storage batteries), and a new industry was born. The Embudo Milling Company erected a grinding plant at Embudo station on the Denver & Rio Grande Western Railroad and shipped over 10,000 tons of lithium-bearing concentrates before suddenly ceasing operations in 1930—due, it is said, to the contamination by some "troublesome" tantalum mineral. The mine ultimately proved to be a veritable treasure trove of rare and unusual minerals (including that very same troublesome columbite-tantalite and microlite, beryl and optical-quality calcite. The Harding pegmatite yielded the lion's share of the nation's tantalum supply—over 20,000 pounds of microlite and columbite-tantalite concentrates during World War II—and at times was the nation's leading producer of beryl. The nearby Iceberg calcite deposit, the second largest ever discovered in the world, produced well over 1,000 pounds of optical-quality material, much of it sold to Bausch & Lomb Optical Company. The U.S. Bureau of Standards, upon examination of the crystals and rhombs, declared them "to be the finest in quality and the largest in size ever known." The mine is now a priceless mineralogical and geological "laboratory" administered by the University of New Mexico, a gift of the late Arthur Montgomery.

The yellowish outcroppings near Questa, originally thought to be sulfur, turned out to be ferrimolybdite and were an indication of a subsurface molybdenum sulfide deposit. The Molybdenum Corporation of America acquired the property from its early owners in 1920 and has since produced molybdenum sulfide more or less continuously. The mine has gone through three distinct phases beginning with traditional underground methods focusing on the narrow, high-grade veins, then evolving into open-pit mining in 1965, and finally back to large-scale underground mining of lower-grade material in 1983. Milling at Questa has similarly gone through many technological changes and improvements since the early days. Questa is truly a world-class deposit, having produced about a quarter billion pounds of molybdenum sulfide to date.

The value of perlite went unrecognized until the late 1940s. Within a few short years of initial testing and research, perlite was found to be invaluable for a whole host of end uses including lightweight aggregates, abrasives, potting soil additive, filtering media, insulation, absorbents, and many others. The No Agua perlite deposit a few miles north of Tres Piedras, currently recognized as the world's largest, gave Taos County its second so-called "world-class" deposit. F. E. Schundler built the first mill in the district and began production in 1951. Johns Manville took over the property in 1959, and since then several other companies and/or successors have been active in the district.

Mica, unlike perlite, has a long history of exploitation in New Mexico but was of little economic importance until recently. One of the largest recently mined deposits of mica (sericite) in the United Stated is at the U.S. Hill mine (aka Tojo) in southeastern Taos County, once owned and operated by the cleverly named "Mineral Industrial Commodities of America, Inc." (M. I. C. A.) and most recently by Oglebay Norton. The ore is mined by standard mechanical stripping techniques, screened, and then trucked to the company's grinding and flotation mill on the Rio Grande near Velarde. Reserves are sufficient for decades of operations, but the future of the mine is in doubt due to very strong resistance of the local populace over environmental and cultural concerns (the local pueblos use the mica in their pottery).

MINING LAWS, THE ENVIRONMENT, AND THE FUTURE

Attitudes and opinions of society have changed radically. Well into the middle of the twentieth century Americans often perceived their local mining activity with a sense of pride. Despite the tradeoffs, such as mine dumps and sometimes malodorous mills, they were keenly aware of the economic benefits. This is the two-edged sword of our mining legacy: The benign side is represented by the vast amounts of wealth and useful materials won from the earth. The negative side is represented by the non-essential byproducts derived from those commodities: tailings dams, slag piles, and other wastes left over from past operations.

What has changed most, however, is the world's population, and that has spawned a staggering demand for the mineral feedstocks of civilization. The mining industry has responded by conducting operations on a much larger and more efficient scale. Where the pick, shovel, sluice, pan, and single-horsepower haulage once prevailed, miners of today take advantage of motorized trucks capable of hauling a hundred times more tonnage in a single load than our pioneer miners handled in an entire week. A similar revolution has taken place in the explosives end of the business: Black powder and hand steel have given way to multi-gang hydraulically operated percussion drills capable of drilling hundreds of feet of blast hole in a single 8-hour shift, which, when loaded with modern-day explosives, break hundreds of thousands of tons in a single blast. The downside of this modernization and efficiency is that piles of non-ore-bearing rock and other waste streams have grown enormously. The magnitude of environmental problems has soared, and we have all become more cognizant of the fragility of our environment.

Our increasing population is rarely perceived as part of the underlying problem; instead our ire is usually focused on the laws that permit the development of mineral deposits in the first place. The days are gone forever when an individual or a company could develop and mine a mineral deposit on and within lands of the United States, whether under the purview of the American Mining Law, the Minerals Leasing Act, or the Saleable Minerals Act, without full environmental compliance. Anyone desiring to participate in such activity today is subject to superior laws and regulations (such as the Clean Air and Water Acts, National Environmental Protection Act, and dozens of others) administered at the federal, state, and local level.

Today, environmental, cultural, and a host of other considerations are carefully weighed against the need to extract resources. If the former are deemed more critical or socially valuable than the latter, a mine such as that at U.S. Hill could quickly become part of history or never make it through the permitting process. These issues will challenge us as we become more, not less, dependant upon a reliable stream of mineral commodities. Balancing our need for them against the desire for a pristine environment and the preservation of our cultural heritage will doubtless be among the more vexing issues of the future, because the earth is not only our home but also our sole source of those resources.

Special thanks to Homer Milford and Spencer Wilson, both of whom have long shared and exchanged historic data and photographs with the author; and to Meghan Jackson for ferreting out information on the "Indian Paint" mines and the local pueblos micaceous pottery.

The History and Operating Practices of Molycorp's Questa Mine

Anne Wagner, *Molycorp, Inc.*

Molycorp's Questa molybdenum mine has been in operation since 1918. For more than eighty years the mine has been a strong source of economic vitality in northern New Mexico. Producing hundreds of millions of tons of molybdenum since its inception, the Questa mine continues to operate, providing a valuable natural resource to the world market while fully committed to operating in a safe and environmentally responsible manner. This paper provides a brief history of Molycorp's operations and insight on the mine's present and future operations.

A HISTORY OF OPERATIONS

In 1914 two local prospectors staked multiple claims in an area of the Sangre de Cristo mountain range called Sulphur Gulch. During their exploration they discovered an unknown dark, metallic material. The common belief at the time was that it was graphite, and it was rumored to have been used for a myriad of functions from lubricating wagon axles to shoe polish. In 1917 a sample of the ore was sent out to be assayed for gold and silver. The report included a mention of molybdenum and its rising value resulting from an increase in usage during World War I. Early in the summer of 1918 the R & S Molybdenum Mining Company began underground mining of the high-grade molybdenum veins in Sulphur Gulch. On June 1, 1920, the Molybdenum Corporation of America was formed and acquired the R & S Molybdenum Mining Company (which later became Molycorp).

By August of 1923 Molycorp had built its own on-site processing mill, which could produce 1 ton of molybdenum concentrate daily from 25 tons of ore. All molybdenum production during this period was from high-grade molybdenite (molybdenum sulfide), with grades running as high as 35 percent molybdenum. This mill was one of the first flotation mills in North America. The mill was rebuilt several times and operated continuously until 1956, when the underground mining operations ceased. In 1963 this mill was dismantled to make way for the current mill.

From 1957 to 1960 exploration by drifting, cross-

Molycorp's earliest workforce, ca. 1920.

cutting, and core drilling methods was conducted under the Defense Minerals Exploration Act. After completion of the contract, Molycorp continued exploration, and in early 1963 core drilling from the surface and underground was accelerated to determine whether or not an open-pit mine was economically feasible. By 1964 sufficient reserves had been blocked out to justify the development of an open-pit mine and the construction of a mill that could handle 10,000 tons per day. Pre-production stripping was started in September 1964, and the first ore from the pit was delivered to the mill in January 1966.

Production from the pit continued until August 1982. During this period the mill capacity increased to 18,000 tons per day, and the stripping rate increased to 120,000 tons. At this time the mine employed approximately five hundred workers. Also at this time, additional exploration drilling in the area of the existing open pit delineated several other orebodies. The largest of the orebodies contained approximately 125 million tons of ore averaging 0.3 percent molybdenum sulfide.

Molycorp was acquired by Union Oil Company of California in August of 1977. In November 1978 development of the existing underground mine began, with two vertical shafts bottoming out at approximately 1,300 feet deep, and a mile-long decline was driven from the existing mill area to the haulage level. The mill flotation area was modernized to accommodate the higher-grade of underground ore. In 1982 mining from the open pit ceased, and in August 1983 the new underground mine began operating. Employment at this time reached approximately nine hundred workers.

In 1986 an extremely "soft" market caused the first shutdown of the mine in recent history. The mine was restarted in 1989 and continued to operate until January 1992, when the mine was shut down again due to low prices. The mine restarted in 1995, and most of that year was devoted to mine dewatering and repair. Production began in late 1996, and over the next several years approximately 30 million pounds of molybdenum concentrate were produced. Development of the current orebody began in 1998. Production from this orebody began in October 2000. This orebody and three adjacent ones collectively have sufficient ore reserves for production to continue for several decades.

MINE OPERATIONS

The mine has operated by several methods over the past eight decades. Initially, the mine was a small, underground working, with donkey-hauled ore cars delivering ore broken up by workers to the surface. In the early 1960s Molycorp decided it was more techni-

Molycorp's molybdenum mill and conveyor system with product in sling bags packaged for sale.

cally feasible to mine the orebody via an open pit.

For twenty years, miners at Molycorp blasted pit rock to access ore. Because open-pit operations generate much more non-ore-containing rock, the company developed a system of disposal through the construction of nine rock piles located throughout the site, common practice throughout the industry. Molycorp disposed of several hundred million tons of rock into these piles. Currently Molycorp is working to ensure the long-term stability of these piles, including the mitigation work on the Goathill North pile that is underway.

In the late 1970s Molycorp determined that the orebody was too deep to mine by an open pit. In 1982, after several years of development at a cost in the hundreds of millions of dollars, the current underground operation was opened. The mine is now an underground gravity block cave mine, designed to produce 18,000 tons of ore per day.

Blocking caving is a bulk mining method. It takes a relatively weak and fractured orebody and collapses it under its own weight. The ore is then drawn down into a series of excavation areas for sizing and transportation by mechanical belt to the surface for further refinement and milling.

During the milling process, the molybdenum ore is physically separated from the rest of the rock. Through the use of a flotation system, the molybdenum is literally "floated" to the top of the tank, and

Molycorp's underground workforce in 2003 standing in lift, ready for transportation into the underground

Molycorp tailings facility interim revegetation. As a routine practice Molycorp caps the inactive tailings areas with soil and seeds with native species.

the remaining material is carried by pipeline nine miles to the tailings facility west of the village of Questa.

OPERATING PRACTICES

Since the mine began in 1918, accepted operational practices have changed. These changes have led Molycorp to adopt and incorporate best management practices throughout its site. Always closely connected with the village of Questa, Molycorp is continuing to work with the village and other stakeholders to promote successful economic opportunities in the community.

Molycorp is using its resources to promote economic development opportunities in two key ways. The first is a comprehensive program to use local contractors whenever possible, and if it is not possible, outside contractors are encouraged to use local suppliers or subcontractors when possible. The second is a program that supports the use of local suppliers and vendors of materials as much as possible. In addition, Molycorp is collaborating with many of the stakeholders who have an interest in economic and social sustainability of the village of Questa.

Environmental practices have also changed over the years as regulations and expectations have changed. Today, Molycorp is devoting significant efforts to environmental investigations, best management practices, mitigation, etc. as the company continues to mine.

Current best management practices along the Red River include collecting poor quality, shallow ground water and treating it to improve the water quality of the Red River. Ongoing voluntary investigations include annual monitoring and reporting on the aquatic biology of the Red River.

On the mine site Molycorp is collaborating with the state of New Mexico, the village of Questa, and other stakeholders to understand the extent of the instability of the Goathill North rock pile and to determine the best approach for long-term stabilization of the rock pile in an expedient manner. Work on the mitigation project began in July 2004 and is expected to be completed in mid-2005.

Another significant effort on the rock piles is an extensive test plot program that was established in late 2003. The purpose of the test plot program is to further refine an appropriate reclamation approach that Molycorp will implement upon closure of the mine. This undertaking exemplifies the uniting of on-site studies, off-site research, and state-of-the-art technology. Twenty-three research test plots were constructed over 20 acres. The test plots consist of areas of various slopes including level areas, and slopes with gradients of 2:1 and 3:1. Various cover depths (zero, one, and three feet) were placed on the plots to evaluate the effectiveness of the cover for reclamation, and more than 26,000 seedlings were planted by hand along with seeding of grass and other plants. It is anticipated that this research will redefine high-elevation reclamation in the western U.S.

THE FUTURE

Although Molycorp anticipates continuing operations into the future, planning for closure is also important. A reclamation plan is required by the New Mexico Mining Act, but Molycorp believes planning for closure goes beyond reclamation. Current best practice in the industry means thinking and planning ten or even twenty years ahead of developments when it comes to closure. Molycorp is now working to shift the community away from its dependency on mining through supporting other economic development opportunities. Molycorp has also committed to continue operating the mine even when prices for molybdenum drop, moving employees from mining to reclamation and remediation projects until prices return to economic levels. Molycorp is committed to mining responsibly, protecting the environment, working safely, and partnering with local communities to develop additional economic opportunities so the communities remain viable into the future.

Metal Deposits in New Mexico—How and Where They Occur

Robert M. North, *Phelps Dodge Chino Mines Company*
Virginia T. McLemore, *New Mexico Bureau of Geology and Mineral Resources*

New Mexico is at the eastern edge of one of the world's great metal-bearing provinces. Copper is particularly important, with large and (in come cases) world-class deposits in Arizona, New Mexico, Utah, and Sonora, Mexico. Nevada and (to a lesser extent) the entire Southwest are well endowed with gold and silver. Some of the largest molybdenum deposits in the world are found in Colorado and New Mexico. This is the nature of metal deposits; they are relatively common in certain geologic settings and almost unknown in others. Prospectors have known this intuitively for many years, and it was the prospector who found most of the important deposits in southwestern North America.

The metals found in New Mexico deposits are often the results of the interaction of hydrothermal (hot water) solutions with a host rock. In some cases further enrichment takes place when cooler, meteoric water (rain and snow melt) interacts with low-grade mineralization to concentrate the metal through a process called *supergene* ("from above") *enrichment*. Many of the deposits discovered so far in New Mexico are the result of water-rock interactions. These same types of reactions can create environmental concerns later on when the deposits are developed.

The important fact is that metals occur in distinct minerals, and to recover the metals, minerals must be concentrated through processing. As higher-grade deposits are depleted, lower-grade deposits will supply the metals that society needs. This is especially important in New Mexico, where low-grade stockpiles and tailings of earlier mining operations exist and could be processed in the future for their metal content. This becomes increasingly significant as technological advances facilitate the economic recovery of metals from lower-grade ores.

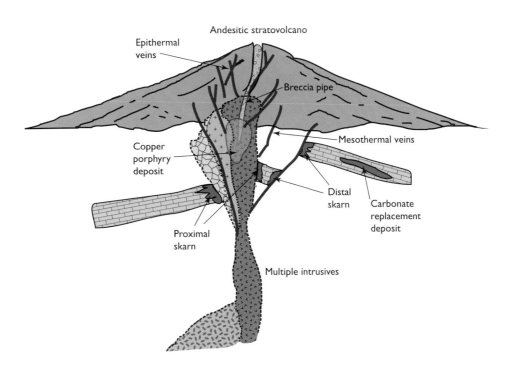

Simplified porphyry copper system. Mineralization typically occurs in and around porphyritic diorite, granodiorite, monzonite, and quartz/monzonite plutons, which are surrounded by concentric zones of hydrothermal alteration. The porphyritic igneous intrusion is locally surrounded by copper and lead/zinc bearing skarns.

CHAPTER ONE

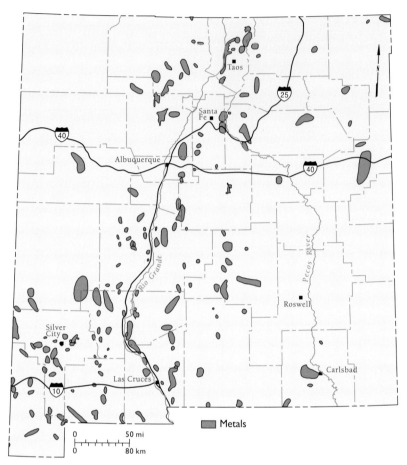

Metal mining districts in New Mexico. Most of these districts are not presently active.

COPPER DEPOSITS

Copper has historically been the most important metal produced in New Mexico. Copper is prized as the second-best conductor of heat and electricity (after silver) and has many uses, especially in the generation and transmission of electricity. Copper occurs in a number of specific geological environments in New Mexico. Most important of these are the porphyry-copper and associated contact metamorphic (or *skarn*) deposits mined in the southwestern part of the state (the Chino mine near Silver City). Sedimentary red-bed type copper deposits are fairly common in New Mexico but have produced only small amounts of copper. Copper is also a byproduct of the mining of zinc and lead vein and volcanogenic massive-sulfide deposits.

Porphyry Copper Deposits

Porphyry copper deposits are the most important metal deposits both in New Mexico and worldwide. These deposits occur in porphyritic igneous rocks and adjacent host rocks of small igneous intrusions at shallow depths within 1.2 miles (2 km) of the surface in the earth's crust. The term porphyritic refers to a specific texture of igneous rocks characterized by large crystals "floating" in a finer-grained matrix, each representing a different episode of cooling. Not all porphyritic igneous rocks host economic mineral deposits, although porphyritic textures are commonly associated with metal deposits. The molten rock or magma intrudes cool host rocks. As the magma cools, it solidifies from the contact toward the center of the intrusion. Volatiles in the magma, primarily steam and carbon dioxide gas, build up within the magma until the pressure fractures the solidified porphyry and host rocks above the magma chamber. Hydrothermal solutions are released through these fractures and react with the host rocks, altering them in a characteristic manner. Copper minerals are deposited as a part of this interaction.

Typically these deposits are very large, some in excess of a billion tons of mineralized rock. Copper grade varies from about 0.10 percent copper to over 1 percent copper with 2–5 percent pyrite. The important deposits in New Mexico at Chino and Tyrone were relatively low-grade ore after this initial mineralization and were upgraded through this process of *supergene* enrichment. The porphyry copper deposits in New Mexico are 55–58 million years old.

Supergene enrichment occurs when rocks with high pyrite content come into contact with water in an oxidizing environment. When this happens, water, oxygen, and pyrite react to form sulfuric acid. This acid naturally leaches copper from the mineral chalcopyrite, putting copper into an acidic solution. Copper stays in solution as long as the solution is acidic and oxidized. However, when the acidic solution is neutralized, or conditions becoming reducing, the copper comes out of solution. Typically in natural systems, conditions become reducing just beneath the water table. When this happens, the copper in solution easily combines with sulfur. The sulfur available in pyrite and the copper in solution combine and replace pyrite with the copper mineral chalcocite. This process preserves copper through chemical dissolution and subsequent enrichment. Otherwise the metal is eventually lost through the erosion of copper-bearing rock by mechanical means. Without this process, deposits such as Tyrone would never have been economic, and the deposits at Chino would be much smaller.

Skarn Deposits

Deposits of copper and other metals can form when hot magma comes into contact with relatively cool, reactive rocks such as limestone, resulting in the distinctive rock termed skarn. The term skarn refers to rocks with diverse origins but similar mineralogy. Commonly these rocks have significant calcium-bearing varieties of garnet and pyroxene, sometimes with significant quantities of magnetite. Although such deposits form in a number of geological environments, they are most common in the southwestern U.S.

Polymetallic Vein Deposits

Polymetallic vein deposits formed as a result of tectonic and igneous activity in the Southwest between 75 and 50 million years ago and are found in a number of districts. These veins exhibit different textures and mineralogy but are similar in form and age. The most important deposits of this type in New Mexico are in the Hillsboro, Pinos Altos, Bayard, and Lordsburg districts. Vein deposits occur where hot, metal-bearing solutions that are formed near an igneous intrusion are channeled through fractures, typically fault zones. These hot-water solutions can deposit metallic minerals when the physical or chemical conditions of the solution change, or through an interaction between the wall rock and solution. For example, something as simple as cooling the solution can cause precipitation of metallic minerals. Boiling the solution is a common mechanism for precipitation of metals. Host rocks can be sedimentary, intrusive igneous, or associated volcanic rocks. Veins were typically worked for both base and precious metals and locally contain uranium, tungsten, tellurium, and beryllium. Mineralogies and metal associations are diverse, even within a district.

The ores from veins of this type have potential for use as quartz-rich smelter flux, which can be an important component of the smelting process. Nearly two million tons of mineralized siliceous flux has been shipped from veins in the Lordsburg district to smelters in New Mexico, Arizona, and Texas, for example. Past production indicates that these vein deposits are small in size relative to porphyry copper deposits.

Typical porphyritic texture, in a quartz monzonite porphyry from the Orogrande district in Otero County, New Mexico. The large white crystals (phenocrysts) are feldspar.

Sedimentary Copper Deposits

Copper is deposited in sedimentary rocks either as part of the sedimentary process or during the process of diagenesis, the compaction and solidification of sediments into rock. The mineralized bodies typically occur as lenses or blankets of disseminated and/or fracture coatings. The copper minerals are predominantly chalcopyrite, chalcocite, malachite, and azurite with local uranium minerals, galena, sphalerite, and barite. Ore minerals in these sedimentary copper deposits are typically associated with organic debris and other carbonaceous material. Sedimentary copper deposits are the world's second most important source of copper after porphyry deposits; however, no large copper-producing deposits of this type are found in New Mexico. The most important sedimentary copper deposits in New Mexico were mined in the Nacimiento district near Cuba and the Pastura district near Santa Rosa.

MOLYBDENUM DEPOSITS

Molybdenum in New Mexico has been produced as a byproduct of copper and uranium mining and is currently produced from primary molybdenum deposits at Questa. Molybdenum is used in steel alloys, and molybdenite, the most common molybdenum mineral, is used as a lubricant.

Porphyry Molybdenum Deposits

Porphyry molybdenum deposits are large, low-grade deposits containing tiny, disseminated veins of molybdenum sulfides and are associated with porphyritic intrusions. They occur in three areas in New Mexico: near Questa, in the Nogal district (Lincoln County), and in the Victorio Mountains (Luna County). The largest deposits are at Questa in Taos County. The deposits consist of thin veinlets, fracture coatings, and disseminations of molybdenite (molybdenum sulfide) in granitic host rock. The deposits are similar in form to the porphyry copper deposits described above but are probably younger, 35–25 million years old.

GOLD AND SILVER DEPOSITS

Gold and silver have been produced in New Mexico mostly as a byproduct of copper mining, but some deposit types were mined primarily for their precious metals. Gold is used as a monetary standard, in jewelry, and in specialized electronic applications. This takes advantage of the fact that gold is malleable, a good conductor of electricity, and resistant to corrosion. Silver is the best conductor of electricity and heat, allowing for specialized applications, as well. Most silver is used in photography, but it is also used in jewelry and electrical components.

Volcanic-Epithermal Deposits

Gold and silver occur in deposits classed as epithermal, a term for deposits formed high in the earth's crust at relatively low temperature (less than 300° C). Epithermal mineral deposits in New Mexico (and elsewhere in the world) are found in structurally complex geologic settings that provide an excellent plumbing system for circulation of hydrothermal fluids. Geologists studying deposits of this type established the now-recognized association between epithermal mineral deposits and active geothermal (or hot spring) systems. However, there are many small hot spring systems with no gold or base metals associated with them. The Mogollon district in Catron County is the most productive example of this type of deposit; other productive examples are the Steeple Rock, Socorro Peak, Chloride, (near Winston), and Cochiti (Sandoval County) districts.

Carbonate-Hosted Silver (manganese, lead) Replacement Deposits

Carbonate-hosted silver (±manganese, lead) replacement deposits formed in southwestern New Mexico approximately 50–20 million years ago. The deposits contain predominantly silver and manganese oxides, in veins and as replacements in carbonate rocks. More than 20 million ounces of silver have been recovered from deposits of this type, but gold is rare. Although silver and manganese are the predominant metals, lead is next in abundance. Manganese has been produced from the Chloride Flat, Lone Mountain, and Lake Valley districts. Economically, the most important silver deposits are in the Lake Valley and Kingston districts.

Great Plains Margin Deposits

Some of the state's largest gold deposits are found along a north-south belt roughly coinciding with the margin between the Great Plains physiographic province and the Basin and Range (Rio Grande rift) and Southern Rocky Mountain provinces. These deposits have similar characteristics that, when considered with their tectonic setting, define a class of mineral deposits referred to as Great Plains margin deposits by workers in New Mexico. Most deposits are associated with igneous intrusive rocks that are 38–23 million years old. The veins have high gold/base metal ratios and typically low silver/gold ratios in contrast with high silver/gold deposits in western New Mexico. The most important deposits of this type are found at Elizabethtown–Baldy, Old Placers, White Oaks, and Orogrande districts.

Placer Gold Deposits

Placer gold deposits were an important source of gold in New Mexico before 1902, but placer production since 1902 has been minor. The earliest reports of placer mining were in the 1600s along the northern Rio Grande. In 1828 large placer deposits were found in the Ortiz Mountains in Santa Fe County (Old Placers district), which began one of the earliest gold rushes in the western United States. Most placer deposits in New Mexico had been discovered by 1900. It is estimated that 662,000 ounces of gold have

COMMODITY	YEARS OF PRODUCTION	ESTIMATED QUANTITY OF PRODUCTION	ESTIMATED CUMULATIVE VALUE ($)	QUANTITY IN 2002	VALUE IN 2002 ($)	RANKING IN U.S. IN 2002
Copper	1804–pres.	>10.2 million tons	>13.5 billion	277,094 tons	419 million	3
Uranium	1948–2001	>347 million pounds	>4.7 billion	264,057 pounds	2.6 million	—
Gold	1848–2002	>3.2 million troy ounces	>394 million	32,337 troy ounces	9.5 million	—
Zinc	1903–1991	>1.51 million tons	>338 million	—	—	—
Silver	1848–1998	>117 million troy ounces	>252 million	433,596 troy ounces	2.4 million	—
Molybdenum	1931–pres.	>146 million pounds	>214 million	7,339,117 pounds	24 million	5
Lead	1883–1992	>367,000 tons	>56.8 million	—	—	—
Iron	1883–1962	8.2 million long tons	17.3 million	—	—	—
Manganese	1883–1963	>1.9 million tons	5 million	—	—	—
Tungsten	1907–1956	1.1 million pounds	250,000	—	—	—
Total	—	—	20 billion	—	475.5 million	—

Estimated total production of major metal resources in New Mexico. Figures are subject to change as more data are obtained (these are conservative estimates). Tons = short tons, unless otherwise stated.

been produced from placer deposits throughout New Mexico between 1828 and 1991. Only four districts have each yielded more than 100,000 ounces of placer gold: Elizabethtown–Baldy, Hillsboro, Old Placers, and New Placers.

LEAD AND ZINC DEPOSITS

Lead and zinc commonly occur together and have been produced in several districts in New Mexico. Zinc has been the more important of the two; approximately 1.5 million tons worth more than $338 million have been produced. Zinc's most important use is in galvanizing of steel; it is also important in the manufacture of brass and other alloys. Lead is used primarily in lead-acid batteries.

Volcanogenic Massive-Sulfide Deposits

Volcanogenic massive-sulfide deposits are polymetallic, stratabound deposits formed contemporaneously with submarine volcanism by hot saline brines. Modern analogies to this type of mineralization include the "black smoker" deposits at ocean rifts. Volcanogenic massive-sulfide deposits contain varying amounts of base and precious metals and are associated with rocks formed between 1,650 and 1,600 million years ago, or perhaps earlier. Four districts in New Mexico contain known deposits of this type: Willow Creek, Santa Fe, Rociada, and La Virgen. Only the Willow Creek district near Pecos has yielded any significant production. Many of the volcanogenic massive-sulfide deposits and suspected deposits are found within or near wilderness areas and probably will not be developed in the foreseeable future.

Carbonate-Hosted Lead/Zinc Replacement Deposits

Carbonate-hosted lead/zinc replacement deposits are found in southwestern New Mexico and formed approximately 50–20 million years ago. The deposits include replacements and minor veins in carbonate rocks. They are typically lead/zinc dominant, with byproduct copper, silver, and gold. The near-surface oxidized zones were typically the most productive in the past. Economically, the most important deposits are in the Magdalena, Victorio (west of Deming), and Organ Mountains districts.

Polymetallic Vein Deposits

Polymetallic vein deposits with important lead and zinc production are found in a number of districts, particularly near Bayard. These veins exhibit different textures and mineralogy but are similar in form and age. The Bayard deposits are the only important lead-zinc producers of this type in New Mexico.

Industrial Minerals in New Mexico—Different from Metals, but Crucial and Complex

Peter Harben, *Las Cruces, New Mexico*

Industrial minerals have been called the third world of the extractive industry. They seem to trail the invariably glamorous metals and the politically charged energy products. Metals have defined industrial ages, and energy has framed government policy. They are obvious in everyday life as steel automobiles and aluminum beverage cans, not to mention utility bills and gasoline prices. What remains in the extractive industry are the more than fifty diverse nonmetallic and non-fuel materials that tend to vanish into common products. Few people at a dinner party think of a bottle of wine as a melted and shaped blend of silica sand, feldspar, limestone, and soda ash encapsulating a liquid based on a fertilized crop and clarified with bentonite, diatomite, and/or perlite; secured by a plastic "cork" filled with calcium carbonate and talc; and identified by a label of cellulose containing kaolin, calcium carbonate, and titanium dioxide.

Despite such a low profile, industrial minerals are essential for our modern society and contribute significantly to New Mexico's economy. New Mexico ranks first in the country in the production of potash, perlite, and zeolites; second in humate; third in copper and pumice/pumicite; and fifth in molybdenum. Our state is also a significant producer of construction sand and gravel and crushed stone (10 million tons per year valued at some $64 million), gypsum, and dimension stone.

Today the total value of non-fuel mineral production in the state of New Mexico is almost $600 million, of which industrial minerals constitute approximately two-thirds. These industrial minerals have a positive effect on society. Potash from Carlsbad is a key ingredient in fertilizer, and salt helps to de-ice roads; travertine from Belen provides unique architectural features; zeolites in Sierra County promote animal hygiene and environmental cleanliness; perlite from Socorro and Tres Piedras contributes to ceiling tiles and horticultural soils; and pumice from the Jemez Mountains and the Mogollon-Datil volcanic field puts the stone in stone-washed jeans. Local limestone from around the state helps remove pollutants from the environment, and construction aggregates from New Mexican limestone, granite, sandstone, and volcanic rocks support the construction boom and improve the state's expanding infrastructure.

This important sector of the economy may represent a viable development path for the future and help to offset declining metal production. What is needed is a clear understanding of the differences between industrial minerals and metals; horror stories abound about metal producers who wandered into non-metals using a metal-mining mentality and were doomed to failure.

LOW PRICES—LOW PROFILE

One prime reason for this lackluster appreciation of the industrial minerals business is the relatively low selling price, with most in the $10–$100 per ton category, though industrial diamonds, iodine, and some rare earths sell for more than $10,000 per ton. What makes industrial minerals interesting is the large number of tons produced and sold. An average price of $5–6 per ton for aggregates in the United States is unimpressive until it is multiplied by the annual production rate of 2.5 billion tons generating sales of more than $14 billion. To put metals in perspective: The value of all non-fuel minerals and metals sold in the United States is approximately $38 billion, of which

The U.S. Hill mica mine in Taos County.

THE PHYSICAL AND HISTORICAL FRAMEWORK

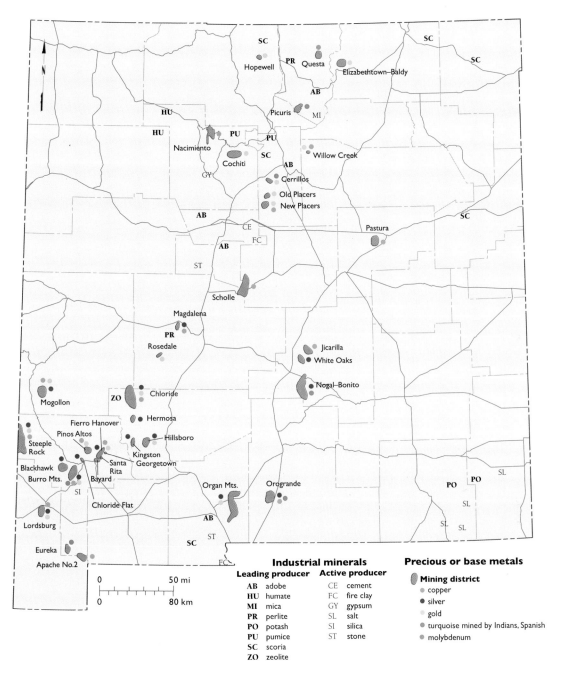

Industrial minerals in New Mexico. Metals are also shown.

$30 billion is for industrial minerals (half of which are aggregates) and $8 billion is for metals. The value of cement sold in the United States is about equal to the *combined* value of copper, gold, and iron ore. Nevertheless, the relatively low price per ton dictates whether or not a mineral can be mined commercially and how far it can be shipped. Consequently, distance to market is more important for the average industrial mineral than for a metal.

LOCATION, LOCATION, AND LOCATION

To borrow a realtors' phrase, three of the most important characteristics of most industrial mineral deposits are location, location, and location. Specifically, certain industrial minerals like silica sand and limestone for glass or ceramics are relatively common and inexpensive and, therefore, restricted to servicing local markets; no local glass or ceramics plant means no market. Less common minerals like perlite or potash

CHAPTER ONE

$1–10/METRIC TONS	
aggregates	
anhydrite	
gypsum	
salt	
sand & gravel	

$10–100/METRIC TONS	
aplite	magnesite
aragonite	nepheline syenite
bentonite	olivine
calcium carbonate	phosphates
cement	pumice
celestite	propylite
dolomite	salt
feldspar	salt cake
ilmenite	zeolites
kaolin	

$100–1,000/METRIC TONS	
alumina	magnesite
anatase	mica
andalusite	monazite
attapulgite	nepheline syenite
barite	nitrates
bauxite	perlite
bentonite	petalite
borax	rare earth oxides
calcium carbonate	rutile
chromite	silica
diatomite	sillimanite
flint clay	spodumene
fluorspar	sulfur
garnet	talc
graphite	vermiculite
iron oxide	wollastonite
kaolin	xenotime
kyanite	zircon

$1,000–10,000/METRIC TONS	
antimony oxide	graphite
asbestos	rare earth oxides
bastnaesite	rutile
bromine	synthetic zeolites

OVER $10,000/METRIC TONS
diamonds
iodine
rare earth oxides

Industrial minerals by price.

can serve regional and even international markets because the higher prices can absorb the cost of transportation, and competition may be more distant or less intense. Still rarer products like borates have limited sources of supply and can service international markets. This need for proximity to the market is of particular relevance to New Mexico, where market demand is limited by a relatively modest regional population and manufacturing base. For the most part, mineral use in North America is concentrated in the traditional markets of the Northeast and Midwest and (more recently) the fast-growing southern states. Manufacturing in the West tends to be restricted and concentrated and serves a relatively large area, thus limiting viable mineral supply points. At the same time, many manufacturing plants are moving offshore. For example, ceramics plants are moving from the United States to Mexico and Asia.

The geographic spread of industrial minerals is being extended through the increased use of elaborate processing techniques to develop value-added grades. Kaolin, for example, may be ground and air floated to produce a low-cost filler used locally for rubber; water washed for use nationally as a fine, white filler in paper; or floated, delaminated, calcined, and magnetically separated to produce a bright "engineered material" for use in lightweight coated grades of paper all over the world. In contrast, perlite is shipped unprocessed in its compact form and later expanded to a much larger volume close to the point of consumption.

SPECIFICATIONS, SOPHISTICATION, AND SERVICE

One key difference of industrial minerals is that they are prized for a combination of several physical and chemical properties, whereas metals are generally valued for a specific element that is extracted from the ore. Specifications for industrial minerals, therefore, may require a combination of physical characteristics such as whiteness, particle shape, size and size distribution, viscosity, oil absorption, as well as chemical purity and strict maximums on impurities such as iron or heavy metals. Each end use requires a different set of specifications; in fact, industrial minerals are often described by the end-use grade—for example, drilling-mud-grade barite, coating-grade kaolin, and refractory-grade bauxite. The importance of the mineral's end use is illustrated by the fact that titanium minerals are included in the industrial minerals category because 95 percent of the consumption is as an oxide for pigment rather than as a metal for jet fighters or bicycles. Other metallics overlap into the industrial minerals camp because of their nonmetallic uses—refractory, chemical, and abrasive-grade bauxite; refractory, chemical, and pigment-grade chromite; beryllium as beryllium oxide; etc.

In contrast, metal specifications are relatively simple, and there is no "fingerprint." Copper from New

Fresh cut at the No Agua perlite mine in Taos County, owned and operated by Harborlite Corporation. Note the stratified layers of volcanic ash that are characteristic of the deposit.

Mexico works as well as copper from Zambia or Chile, and 14-carat gold from a family heirloom sells for the same price as freshly poured bullion from Witwatersrand, South Africa. Place value is relatively unimportant in metals. Therefore, these undifferentiated metals command little or no consumer loyalty and are sold at published prices. In contrast, many industrial minerals need to be characterized by a multitude of chemical and physical characteristics and often require an exhaustive evaluation process in order to qualify for use by a particular consumer or plant, which may be optimized for material from a single mine. This requirement tends to generate loyalty to the supplier, and this loyalty may be cemented further as the supplier works with the consumer to improve quality or delivery methods, reduce transportation costs, develop new grades, etc.

In some cases, specifications can only be met by blending minerals from various sources. For example, graphite houses may blend graphite from China, Brazil, and Canada, and a large paper plant may work with a blend of kaolin and ground calcium carbonate. Suppliers of industrial minerals tend to encourage customer loyalty through good service, consistent quality and delivery, and even joint technical research. This makes market penetration for a newcomer difficult, because a lower price is not always a means to wrest away a market share. To make matters still more complex, the selling price of industrial minerals is reached through negotiation between supplier and consumer, based on the quantity purchased and the terms of the contract, rather than (say) the fixed market price of gold set daily in London.

If an industrial mineral fails to meet the industry specifications on a consistent basis, then it is unlikely to be sold regardless of price. In contrast, metals such as gold or copper can be mined and refined with the sure knowledge that they can be sold at a price approaching a level dictated by a terminal market and published in the press or on the internet. In the case of products like iron ore, the buyer can financially penalize the seller for higher levels of impurities such as phosphorus or arsenic. This rarely happens with industrial minerals; the product is simply rejected.

This increased sophistication has been encouraged by the advent of fast, automated methods of manufacture demanding raw materials with a high degree of uniformity. This in turn has placed heavy responsibilities on the suppliers of raw materials and has reduced still further the concept of "price is king" as it is in the metal markets. Today in the industrial minerals arena it is price plus quality, consistency, and service that makes for a complex and demanding business.

WITHOUT A MARKET, A DEPOSIT IS MERELY A GEOLOGIC CURIOSITY

Metals and nonmetals share some common ground—both require geologic exploration, mining, materials handling, and processing designed to produce commercially viable materials. Even here there are exceptions, as many large industrial mineral deposits are obvious surface outcrops or may be discovered incidentally during the process of drilling for oil, gas, or

The Harding pegmatite mine in Taos County during its productive years.

Split-face travertine at New Mexico Travertine in Belen.

water. Consequently, high-profile regional exploration programs for common industrial minerals are rather rare. More commonly, the many known industrial mineral deposits are continually evaluated based on location, transportation, quality, processibility, and market demands.

Industrial minerals are market driven, a fact encapsulated in the saying that "without a market, a deposit is merely a geologic curiosity." This means that a detailed market study is required to test the economics of a given project. However, most modern mining companies employ a crew of managers, accountants, lawyers, investor relation gurus, engineers of all stripes, plus the odd geologist and even botanist, yet very few feel the need to engage in-house market research help. At the same time, companies continue to build mines and plants on the fly, and exploration geologists still scour the far corners of the world for industrial mineral deposits without the slightest notion of whether anything found can be sold at a profit or even sold at all. The concept that industrial minerals demand strong market analysis in order to become viable and valuable is lost in the rush to drill, raise capital, engineer, construct, debug, and produce a product on schedule.

There are important differences between the economic behavior of industrial minerals and metals based on their different end-use pattern, which in turn dictates the method of marketing required. In general, industrial minerals tend to be used more in the non-durable consumer-goods sector of the economy, whereas metals are more in the durables and engineering sectors such as transportation. These differences in end-use markets are reflected in variations in demand behavior. Capital cycles that affect metals tend to be more extreme than consumption cycles for industrial minerals, which tend to lag behind. As consumer demand for industrial minerals advances, it generates pressures for the creation of new production capacity. The markets for industrial minerals are regarded as "steady" whereas metal markets tend to be more volatile (however, when copper prices climb above $1.00 per pound, industrial minerals become "boring" rather than "steady"). Some mining companies find that the steadiness of industrial minerals—cat litter is still necessary during a recession, but a new washing machine can be delayed—is a good counterpoint to the volatility of metals.

CHAPTER TWO

ENVIRONMENTAL AND WATER QUALITY ISSUES

DECISION-MAKERS
FIELD CONFERENCE 2005
Taos Region

Rio Grande in fall color, near Embudo.

Public Perspectives on Mining—"There Must Be a Way to Do It Right"

Stephen D'Esposito, EARTHWORKS

Mining operations of one type or another have been altering our landscape since the dawn of civilization. Today there is growing evidence that public opinion and the political landscape within which mining operates is altering as well. This is due primarily to conflicts over land use and public values.

> Pollution…air, water and land pollution-it runs the gamut…the destruction of beautiful mountains and valleys…the Berkeley Pit is one of the biggest toxic waste sites in the country…cyanide…jobs…natural resources…helped build the country…destroys the countryside…its better than it used to be…they haven't figured out a way yet to deal with the harmful byproducts and keep the land healthy…what do we have to show for it…what's it going to do for us once the mine is gone and they've ruined the environment and the jobs are gone…employs a lot of people and I support it but I wouldn't want to live near a mine…there must be a way to do it right.

These excerpts from focus groups commissioned by EARTHWORKS in 2003 in Billings, Montana, Las Vegas, Nevada, and Seattle, Washington, offer a sampling of public attitudes about mining today. And they match the views expressed in focus groups and polling that we sponsored in 2000, with one significant difference—although mining remains a remote activity for most, there is a growing awareness of its impacts particularly in states with significant current or historical mining activity. The participants in the 2003 focus groups were more able to cite specific mining accidents and even specific types of pollution such as cyanide spills or acid mine drainage.

There is a difference between public awareness in mining regions as opposed to those regions where there is little large-scale mining. For example, people from non-mining urban areas like New York, Baltimore, and Chicago have a vague but somewhat negative impression of mining. Their views are shaped to a great extent by historical images of coal mining. People in the West, even the urban West, have often seen mines, have family members or friends who work in the mining industry, and can draw upon images (from both actual experiences and media stories) of mining and mining accidents and pollution. Their understanding of the role that mining has played economically, either positive or negative, typically comes directly from experience.

The generally negative perception of mining is balanced by an almost unanimous desire for mining, when it occurs, to be done the right way. The public is not anti-mining and is well aware of the material and economic needs that are met by mining the earth's crust. People understand that they benefit from the products created by mined materials, and they want to continue to enjoy those benefits but as one focus group respondent stated: "There must be a way to do it right." This sentiment was shared by virtually every focus group member in every focus group that we have conducted. The public wants tougher mining safeguards, and they want greater corporate responsibility-in fact they expect it.

TOUGHER STANDARDS AND INDEPENDENT VERIFICATION ENHANCE PUBLIC CREDIBILITY

Many in the mining business acknowledge that the industry has a public-perception problem; some even acknowledge that they have a reality problem. In 2000 a number of major mining companies sponsored a multi-year study known as the Mining Minerals and Sustainable Development (MMSD) project and launched a major effort to reposition the industry, including the creation of a new global trade association with a mandate to address issues related to sustainable development—the International Council on Mining and Metals (ICMM). There is, however, a split in the industry as to whether improving public perception is a matter of better performance or just better public relations. Those who look at the issues seriously recognize that better performance is essential to a better reputation. They are therefore focusing their time, energy, and resources on practices and performance first, rather than public relations. They also recognize that an essential component of a better reputation is an acknowledgment of improvements from those outside the mining industry—e.g., non-governmental organizations (NGOs), regulators, academics, investors, insurers, and those that market and sell

Placer Dome's Golden Sunlight gold mine near Whitehall, Montana.

products that use metals such as jewelry retailers and high-tech companies. Many realize that the public is likely to see the cues that come from these groups outside the mining industry as more credible than those that come directly from the mining industry or from mining industry-sponsored groups or trade associations. The public tends to discount self-promotion, as it should. It is always better to have someone else pat you on the back or offer kudos.

Interestingly, the two ingredients that are essential to improving the practice, performance and reputation of the mining sector—stronger government regulations and mechanisms for independent, third-party verification of responsible practices—are fiercely controversial within the mining sector.

In regard to stronger mining laws and regulations, a number of mining companies have asserted to me that they are better corporate citizens but they are not getting credit for it. In the U.S. a number of the major mining companies appear to have become smarter about where they propose mines. Yet, some companies still propose mines near national parks, under Wilderness areas, or in important watersheds. As a result, the reputation of the entire industry is tarnished. Laws or regulations that block irresponsible companies from proposing mines in these areas actually enhance the reputation of the industry and allow companies to focus investment resources for mines in more suitable areas.

In regard to independent certification of performance, instead of movement toward independent standards and verification, there appears to be a preference in the mining industry for self-imposed, first-party standards regarding environmental and social performance. This strategy lacks transparency, credibility, independence, and legitimacy. It would be like Pepsi conducting a taste test to compare itself to Coca-Cola, without blindfolds and using its own employees. Fortunately, there is recent evidence that some leaders in the industry may be willing to participate in the development of an independent, third-party verification scheme, with standards developed through a transparent process and with participants from multiple sectors. This is a potentially significant development.

It is true that the reputation of the entire industry suffers because of the "bad apples," a point made on numerous occasions by leaders in the mining industry. Stronger mining regulations and independent verification schemes could begin to differentiate publicly the leaders from the laggards—and should provide *reputational*, financial, and regulatory rewards to leaders in the industry. Such an approach is a risky strategy only for the laggards. It is a shrewd strategy for those who consider themselves to be industry leaders and are willing to prove it. Those who pursue this strategy may be less popular at trade association sponsored cocktail parties but are likely to find themselves more highly regarded by investors, insurers, and the public.

VALUES VS. TECHNICAL ARGUMENTS ABOUT HOT BUTTON ISSUES

There are today a number of hot-button issues that tend to foster public controversy and push the reputation of mining into the negative column. I will describe two such issues: 1) mine location and land-use conflicts and 2) responsibility for closure and cleanup and the definition of responsible reclamation. In each of these examples the public is listening to arguments that are both technical and value-laden, but the industry is typically making only technical arguments. Mining company officials too often act as if the less-quantifiable values held by the public are irrelevant.

Mine Location and Land-Use Conflicts

Modern industrial-scale mining can alter landscapes, water systems, economies, and communities, often in ways that are permanent or long-lasting. Therefore, it should come as no surprise that mining proposals are highly controversial when they conflict with other land uses or preservation.

Near Sandpoint, Idaho, a mining company wants to mine under the Cabinet Mountain Wilderness Area and they want to place the tailings in an unlined pile near the Clark Fork River, threatening clean water, the scenic beauty of the area, endangered grizzly bears, and trout populations. Most people in Sandpoint, and a vast majority of business leaders, oppose the mine; they do not believe this is an appropriate place for a mine or for the mine waste that would result.

However, because of the nature of the laws and the regulatory process, the policy-making process fails to adequately account for the key public concern—the suitability of the location and the fact that the public values keeping the land, and its wildlife and other natural resources, in a protected state. Instead, the debate takes place through competing scientific and technical studies regarding the potential impacts of the mine. These studies are essential, but they fail to address the underlying question of land suitability. The mechanisms that do offer some protection, such as land withdrawals, are not always adequate or effective. Some public officials have had to resort to expensive, complex, and messy buyouts and land exchanges to prevent mine development. What's lacking is a process that allows for the effective and efficient weighing of land uses that should be at the heart of good land use planning and decision making.

Until we begin to develop standards and norms for making appropriate land use decisions, standards and norms that are also accepted by nearby communities, civil society groups, mining companies, and other stakeholders, these conflicts will continue, and the industry as a whole will be swimming against the tide of public opinion and public values.

Defining and Paying for Closure and Cleanup

A prevalent theme in focus groups, particularly in western cities, is that the mining industry has a reputation for walking away from mine cleanup liabilities—to oversimplify, as one focus group participant put it: "They get the gold, we get the pollution mess and the bill for cleaning up the mine."

There is a historical component to the problem. Some of the worst sites on the Superfund National Priorities List are mines. Then there are abandoned mines that are not part of the Superfund program. Several studies, using different definitions of an abandoned mine, have arrived at different estimates for the number of such sites. For example, the Western Governors Association has estimated that there are 250,000 abandoned mines in just the western states, but limited information from some states means the number could be higher. Using a different definition, Mineral Policy Center (now part of EARTHWORKS) has estimated that there may be as many as 500,000 abandoned mines across the entire country. The point is not the debate over the exact number, it's that the problem is significant, particularly as more and more people begin to live or travel near these sites. There is also a contemporary aspect to the problem. In Montana just a few years ago Pegasus Gold walked away from Zortman-Landusky and other mines leaving taxpayers with a bill of at least $30 million. In March 2003 Jim Kuipers authored a report that showed a potential $12 billion gap—the difference between existing financial guarantees and what was likely to be necessary for adequate mine closure—at operating mines in western states.

In the state of New Mexico it took years of citizen-generated pressure and dogged work by regulators to require the mining companies running some of the state's biggest copper mines to update and increase their reclamation bonds so that an adequate financial guarantee would be in place. Mining companies used technical, procedural, and economic arguments to delay complying with New Mexico law requiring them to post an adequate mine reclamation bond. In a narrow sense, these companies may have benefited in that they delayed posting the bond or perhaps decreased the bond amount through torturous negotiations and delays. But in a larger sense the reputation of the min-

Open-pit mine.

Acid mine drainage in Fisher Creek, Montana, from abandoned mine workings.

ing industry suffers. A good part of the public and the groups representing nearby communities in these struggles now believe that all big mining companies will fight paying to clean up the mess that they have made. At its core this is an issue of public trust—as one participant put it: "It's about cleaning up the mess you made; we all learned this value as children."

SCIENCE AS AN ALLY OF PUBLIC VALUES

Drawing a distinction between issues that lend themselves to technical analysis and those that do not is not an effort to paint science as an enemy of good, sound decision making on natural resource issues. Science is essential to good decision making. The argument, however, is for the need to rely on *both* science and values in weighing public policy issues.

There is a compelling need for more independent scientific research and evaluation in the mining sector.

Too many of today's mines were approved based upon environmental impact statements that under-predicted impacts on water resources. Sound public policy and sound natural resource planning require sound analysis of the expected impacts—i.e., accurate predictions. Without accurate predictions, the public and regulators are being asked to make decisions in an information vacuum. We now have a data set for modern industrial-scale mines in the western states that spans decades. We should study the data set and learn from it. And that's what we are doing. EARTHWORKS has commissioned an independent analysis of environmental impact statements from major mines in the western U.S., and the predictive models that underpin these assessments. We expect to have results in early 2005.

Accurate predictions will enhance the public image and credibility of the mining industry. It will also increase the confidence of investors, insurers, and regulators. After all, it's not just the environment that suffers if impacts are worse than predicted. The result can be dramatically increased cleanup and closure costs, greater financial liability, costs associated with legal action, government intervention, and a negative impact on corporate reputations. Environmental professionals in mining companies, in the consulting sector, and in academia have a particularly important, and potentially catalytic, role to play in solving the problems related to under-predicting impacts. If they do not address these problems (if they rely upon incomplete or inaccurate predictions), they run the risk of being tarred with the same brush as some of the worst companies in the mining industry.

THE PUBLIC WANTS SOLUTIONS

Ultimately, the public is looking for solutions. What they care about is not mining per se, but the products, resources, and way of life that mining makes possible—e.g., a wedding ring, computers and cell phones, and fuel (whether for their own transportation or to get things to them like food, clothes, books, etc.). They want things that are essential as well as things that are not so essential. They also want to protect people, communities, and the environment. They desire to have these products without doing harm, and they desperately want to believe this is possible. They want to believe that corporations can act responsibly. But they know that without accountability some pretty nasty things can happen. They are willing to make choices as consumers, particularly in urban areas, to purchase responsibly sourced products. For example, they want to know how to recycle cell phones, computers, and other similar items, and they want the option of knowing that their wedding ring comes from a responsible source, not one that violates human rights or environmental standards.

There are some positive signs. In Montana the Stillwater Mining Company established a good neighbor agreement with nearby communities, agreeing to enhanced water protection provisions and an extensive water testing program. BHP-Billiton, Falconbridge, and Western Mining Corporation have all agreed not to engage in the practice of riverine tailings disposal at future mines. Recently the Newmont Mining Company decided not to pursue its plan to expand its mining operations to Cerro Quillish in Peru, a sacred mountain and key water source in the region, stating that it had underestimated community opposition. In October, Reuters News Service ran a story describing how BHP-Billiton had worked to gain participation and support from community groups and NGOs like Oxfam for its Tintaya Copper mine in Peru. This past April, the world-renowned jeweler Tiffany & Company called for reforms to outdated mining laws and responsible sourcing of metals such as gold and silver as well as diamonds. And Jewelers of America, the national trade association, publicly expressed support for responsible mining practices.

As a practical way forward, there are four reforms necessary in the mining industry:

- A policy that allows for the designation of valuable lands where mining may not be appropriate and the establishment of a mechanism to efficiently and quickly weigh land-use options. This must be determined early, before significant resources have been spent on development.

- An effective regulatory definition of necessary reclamation of lands and other natural resources with a financial guarantee to cover these costs as a pre-condition of mining. And a fee on current mining should be established to begin to tackle the legacy of polluting abandoned mines.

- A company-specific and then industry-wide commitment to independent third-party certification of responsible mining practices.

- The promotion of recycling, re-use, and smart design to lessen our demand for virgin minerals, in order to fully utilize, conserve, and re-use our valuable natural resources. We should mine metals from cell phones, high-tech equipment, and other products.

As a society we will continue to draw resources from the earth for our survival, well being, and enjoyment. We will continue to dig, probe, quarry, shovel, and drill the earth for its resources, to find new materials and new uses for existing materials. But there is growing evidence that we are beginning to alter our views as to how this should be done and under what conditions. And there are signs that a new ethic of *materials responsibility* is beginning to take hold for communities, the business sector, and governments. Materials responsibility begins where we begin to mine the earth's surface resources. And it begins by asking and answering two questions: What is the most sustainable, most responsible, and most efficient way for society to meet its metals or materials needs, and when we need to mine, what are the standards that should and must be met?

The Environmental Legacy of Mining in New Mexico

James R. Kuipers, *Kuipers and Associates*

The environmental legacy of mining in New Mexico can best be understood by examining human health, water quality, and land use concerns, and regulatory responses that have occurred over time. New Mexico shares approximately the same measure of environmental legacy in terms of historic abandoned mine sites as that of other western states that have similarly undergone extensive exploration and development for minerals over the past 150 years. Today New Mexico has virtually no small to medium-sized hard rock mining industry players, although eight of the 162 major hard rock mines that have operated in the U.S. since 1972 are located in New Mexico. Although these mines only represent less than 5 percent of all hard rock mines in the U.S., it is significant that the attention shown to them as a result of New Mexico's unique environmental regulatory process, both within the state and in the U.S., has made them among the most important mines in the nation, from an environmental legacy standpoint.

HISTORIC MINE IMPACTS

New Mexico is home to possibly the oldest mine in North America in the Cerrillos Hills, where turquoise was first worked by native people at least 1,000 years B.C. It is conceivable that New Mexico is also home to some of the earliest mining-related environmental impacts. The impact of mining in the Cerrillos area is noticeable, and historic mining activities in the region have presented safety issues that attract the attention of the public and regulators today.

In 1820 Josiah Gregg, writing about placer operations south of present-day Santa Fe, noted that in "some places the hills and valleys are literally cut up like a honey-comb." The workings observed by Gregg were created by Spanish miners and differed little from other gold diggings in New Mexico during that time.

In New Mexico, as in most other western states, the primary concern was with jobs and prosperity, and the environmental concerns of today scarcely entered the picture. Prospecting and developing mines were all part of the fulfillment of "manifest destiny" that was so prevalent during the settling of the western U.S. Until the early twentieth century and the coming of larger mines and more obvious impacts, there was little sense of serious environmental problems related to mining. However, in the latter 1800s and early 1900s, aided by the development of modern technologies including nitroglycerin-based explosives and processing techniques, New Mexico's mining industry turned from near-surface placer mining to underground hard rock mining-for silver, gold, lead, zinc, copper, and other minerals. In many cases the prosperous mining towns of that era are today's abandoned mining towns, with the worst of them destined to become Superfund sites.

The Cleveland mill in Grant County, an abandoned lead, zinc, and copper mine and mill is one such example. Although the mine site itself only occupies about 4 acres, tailings have contaminated an additional 10 acres of the bed of Little Walnut Creek, and runoff from the facility had acidified Little Walnut Creek and contaminated it with metals, which may have affected residential wells. Superfund cleanup activities conducted in the 1990s for the Cleveland mill site included removal of the tailings to a repository, in addition to capping and revegetation.

Other legacies were less costly but still significant. In the Socorro area many mine shafts and adits were left open presenting a hazard to public safety. At the same time a significant population of Townsend's long-eared bats was discovered hibernating in the abandoned mine workings. New Mexico's Abandoned Mine Land Bureau secured the mine openings in a way that left them open for bat use.

More recently, New Mexico's response to environmental concerns has been characterized by a slowness to adopt regulations, followed by progress, but accompanied by a failure to address the issue at the state's largest mines. However, in the past five years that obstacle has been removed, although with significant compromise. The following discussions clarify some of the history that has led us to this point as well as where we might go in the future.

Uranium and Human Health Impacts

New Mexico was the scene of some of the earliest vanadium (and later uranium) mining in 1918 near

Shiprock. Despite extensive documentation from European uranium-radium mining districts and from radium watch-dial painters about the health effects among miners and workers who were exposed to uranium decay products, work place protections were not adequate to protect miners and millers. By the early 1960s evidence began to mount of lung cancer among Navajo, Anglo, and Hispanic miners. Following extensive litigation against mine operators and the U.S. government demanding compensation for miners who ended up with lung cancer and other diseases caused by exposure to radioactive materials, the U.S. government agreed to establish a compensation program, acknowledging that workers had not been adequately informed about or protected from known risks at uranium mines operated before 1971.

Since that time New Mexico has served as an example for major changes in mining industry practices in terms of occupational safety and health, and for a unique national compensation program. In addition, New Mexico has developed and demonstrated uranium mill tailings remediation techniques. Although many concerns surrounding this very tragic situation have yet to be satisfactorily resolved, the state has learned a great deal from this issue.

HARD ROCK MINES AND WATER QUALITY IMPACTS

Hard rock mining environmental issues in New Mexico are most often identified with the state's largest hard rock mines-specifically the Questa mine in north-central New Mexico and the Tyrone and Chino mines in southwestern New Mexico. Although very different in terms of metal mined and geography, they have all received their share of attention from the public interest community and individuals who feel they, or their environment, have been affected by the mining operations.

Questa Mine

The Questa mine, now owned by Molycorp, began production in 1918 of molybdenum disulfide ore from underground mines. Open pit development began in 1964 and continued until 1982 by which time some 350 million tons of overburden had been removed and 81 million tons of ore had been mined. The overburden and waste rock was deposited onto steep mountain slopes and into tributary valleys immediately above and along the Red River. In 1983 the mine went back underground using block caving methods.

The resulting impact on the environment has led to one of the most complex and therefore difficult mine remediation projects in North America, and arguably the world. The mine site is characterized by an abundance of acid-generating rock and minerals containing toxic metals located in an environment that ranges from semiarid to relatively wet at the upper elevations. The mine site contains waste rock piles that are nearly 3,000 feet high at elevations that range from 6,000 feet to more than 10,000 feet.

The Red River is located directly adjacent to the waste rock piles and down gradient from the tailings impoundments allowing easy conveyance of pollutants from the mine waste and tailings to the Red River and ground water supplies. Over 3,000 acres of land have been disturbed by mining operations at the Questa mine. Scientific studies have confirmed that significant water quality impacts can occur as a result of contaminated ground water emanating from the mine and mill sites. In addition, steep slopes at the mine site have been determined to be geotechnically unstable with the potential for catastrophic failures. Litigation addressing residents' concern for more than one hundred tailings pipeline spills and tailings site seepage impacts has been instrumental in the evolution of Molycorp environmental management practices and New Mexico environmental programs.

Chino and Tyrone Mines

Lore has it that the Chino mine was first revealed by an Apache Chief to the Spanish in the early 1800s as a turquoise mine. It is now owned by Phelps Dodge Corporation and has become one of the largest copper mines in North America. Some 9,000 acres (more than 14 square miles) have been disturbed by the

The Santa Rita pit and north stockpile at the Chino mine.

open pit mine and associated waste rock piles, leach piles, and tailings impoundments. The site is characterized by a preponderance of acid-generating rock and minerals containing toxic metals, and although located in a semiarid to arid environment, has resulted in the release of significant contamination to ground and surface water. Many tailings spills have occurred in Whitewater Creek, and ground water contamination is evident throughout the mine and tailings sites.

The Tyrone mine, which was not exploited until relatively recently, is also a massive open pit copper mine

The Tyrone mine's numerous open pits, waste rock, and leach piles.

that has resulted in more than 6,000 acres of disturbance. It is characterized by acid generation and the potential for toxic metals leaching from its many open pits, waste rock piles, leach piles, and tailings impoundments. In addition to prevalent ground water contamination, the mine was recently cited for bird deaths at the tailings facility, leading to concern over the potential for mortality to the southwestern willow flycatcher, an endangered species.

Other Mines

Many other mines in the state have been noted for their environmental impacts, including:

- The Cleveland mill site near Silver City
- The Continental pit near Bayard
- The Copper Flat mine west of Truth or Consequences
- The Cunningham gold mine near Santa Fe
- The Asarco Deming mill in Deming
- The Pecos lead-zinc mine and mill complex in San Miguel County
- The ill-fated Earth Sciences, Inc. copper mine near Cuba

More than thirty-five uranium mines and seven mills—two of which were designated Superfund sites—are located between Laguna Pueblo and Shiprock, and the largest potash mining district in the U.S. is in southeastern New Mexico. These mines have been noted for a variety of ground and surface water impacts. In addition, mining often involves the dewatering of large areas, which can create long-term ground water deficits and affect surface water flows as well as seeps and springs.

MINING ENVIRONMENTAL REGULATION

The issue of mining environmental impacts and associated post-mining land use was recognized beginning in the 1960s and had been addressed, in various forms, by most states by the late 1980s. As late as 1993 New Mexico was one of two western states without specific legislation to address either reclamation and closure of hard rock and other non-coal mines or financial assurance to ensure that environmental problems would be adequately addressed at the company's expense. (The other state without such legislation was Arizona.)

However, the problems were recognized by concerned citizens in mining communities, public interest organizations, state government, and even the mining industry itself. The various groups worked together actively during more than three years of negotiations to address the problems The result in 1993 was the passage of the New Mexico Mining Act by the state legislature. This act provides a powerful model of effective mine reclamation policy.

CREATION OF COMPREHENSIVE STATE POLICY

Together with the New Mexico Water Quality Act, the New Mexico Mining Act provides a comprehensive regulatory process to address environmental impacts from mining activities. It does so through a commingling of the authorities of two state agencies: the Environment Department (enforcing the Water Quality Act) and the Mining and Minerals Division of the Energy, Minerals and Natural Resources Department (enforcing the Mining Act). In the end, the process requires the applicant to obtain:

- A discharge permit from the state Environment Department for any discharges into ground water

- A discharge permit (known as National Pollutant Discharge Elimination System or NPDES) from the federal Environmental Protection Agency (New Mexico has not asserted jurisdiction over surface water discharges in the state)

- A financially guaranteed and Environment Department-approved closure plan to assure ground water protection after the end of mining

- A financially guaranteed and state-approved closeout plan from the Mining and Minerals Division

The closeout/closure provisions also require the company to post financial assurance to guarantee the performance of closeout/closure stipulations that address mine reclamation and long-term operations and maintenance.

The enforcement of these laws began with the adoption of New Mexico's innovative Ground Water Protection Regulations, the most significant progress occurring since the passage of the Mining Act. More than one hundred mine sites have been addressed using New Mexico Water Quality and Mining Act regulations, as a result of which significant discharges have been addressed, financial guarantees of post-mining rehabilitation established, and mined land reclaimed, protecting water resources and habitat. The largest mines, facing significant liabilities, lagged behind the rest of the industry, and considerable effort by the agencies, concerned citizens, and public interest groups has been required to get the companies to comply.

Finally, in 2004 nearly all the requirements had been met, with closeout/closure plans and financial assurance being approved for all but a few suspended operations. In doing so, New Mexico has identified the difficulties associated with environmental impacts and reclamation of its mines. The financial assurance amounts for the three largest mines, which are the largest for any mines in the U.S., indicate the enormity of potential environmental impacts. Together they total more than half a billion dollars in potential liability. However, the situation is bittersweet: Although the liability has been recognized, it is mostly covered by corporate guarantees rather than real financial assurance, and the pace of actual reclamation at the largest mine sites still lags behind public expectations.

All seven uranium mills have been reclaimed, the potash facilities remain outside the scope of the Mining Act because of a special exception, and New Mexico has yet to effectively address most inactive and abandoned non-coal mines.

THE ENVIRONMENTAL FUTURE OF MINING IN NEW MEXICO

As New Mexico moves into the twenty-first century, it struggles to reconcile its mining environmental past with its future. A great deal of progress has been made in dealing with the problems recognized in the last century, particularly in the past ten years. It is unclear whether state government and particularly industry have the resolve to face the challenges and to meet the intent of the progressive laws that were passed.

The laws that were intended to protect the state's citizens and resources should provide these citizens the ability to protect their air, water, habitat, and landscapes for the foreseeable future. A critical chapter of New Mexico's environmental legacy of mining is currently being written as mine operators seek "alternative abatement standards" to allow contamination more severe than that allowed by regulation. The citizens of New Mexico have yet to tire in their quest for protection of their water resources and other interests and will continue to work with the present administration to address those issues.

Suggested Reading

Smith, Duane A., 1993, Mining America: the Industry and the Environment, 1800-1980. University Press of Colorado.
Marcus, J., Ed, 1997, Mining Environmental Handbook. Imperial College Press.

Suggested Web Sites

www.amigosbravos.org/molycorpwatch
www.gilaresources.info
www.sric.org

Abandoned Mine Lands in New Mexico

Abandoned mine lands (AML) are a complex issue, not only in New Mexico but throughout the U.S. The following article is intended to provide an introduction to the problem and a look at a number of successful AML programs underway in New Mexico. It was compiled by Greer Price from material provided by Robert Evetts, retired AML Bureau Chief of the Mining and Minerals Division, New Mexico's Energy, Minerals and Natural Resources Department; Virginia McLemore of the New Mexico Bureau of Geology and Mineral Resources; and Melvin Yazzie of the Navajo Abandoned Mine Lands Reclamation Program. Additional information on the Mines Database came from Gretchen Hoffman and Maureen Wilks of the New Mexico Bureau of Geology and Mineral Resources.

Remediation of physical and environmental hazards resulting from inactive or abandoned mines, collectively known as Abandoned Mine Lands (or AML), is one of the greatest challenges to the mining and environmental industries today. Surface and subsurface land disturbance has created serious physical and environmental hazards. Accidental deaths occur every year at old mines across the country. Although some of the more than 500,000 sites in the western U.S. pose little safety or environmental risk, there are many hazardous sites that should be addressed.

Historically miners went about their business much differently than they do today. Exploration commonly involved construction of pits, shafts, and adits that did not yield discoveries and were abandoned. As deposits were found and developed, ore processing facilities (mills, tailings, stockpiles, etc.) were typically located near mine openings. Processing waste was dumped near the processing unit, often into arroyos, rivers, lakes, or other drainages. There were no laws or other guidelines to prevent such contamination. In both underground and surface mining, minimizing costs was a higher priority than safety and stability. Mines generally had few ventilation shafts in order to save money. Upon depletion of the ore, mines were often abandoned and mine waste and tailings piles left as they were, without caps or vegetative cover. Such practices are unacceptable in today's mining industry, but the legacy of these older mines remains.

Although no complete inventory exists, it is estimated that there are more than 15,000 abandoned mine openings (shafts, adits, and pits) in New Mexico with at least 4,000 abandoned mines that have not yet been remediated. Some of these mine features pose significant health and safety issues to the general public. Less than 10 percent of these are coal mines; the vast majority are metal or hard rock sites. Often very little information is available on these mine areas. Many of these sites are on public lands, where they present public land managers with unique challenges related to accessibility and remediation.

PROBLEMS ASSOCIATED WITH AML

The hazardous effects of AML sites on the environment can be broadly divided into the following interrelated and complex categories:

Land Surface Disturbances

Mining by its very nature requires disruption and disturbance of the land surface and/or subsurface. Topography, ore deposit type and shape, economics, and climate all play important roles in determining mining methods and the extent of land surface disturbance. Erosion, sedimentation, subsidence, differential settling of land fills and regraded mine areas, and reshaping of geomorphic features are some of the specific problems that can result. Surface subsidence, such as that occurring above the underground operation at Molycorp's Questa mine, is a natural consequence of some mining and occurs when strata overlying underground workings collapse into mined-out voids, typically as sinkholes or troughs.

Safety

In addition to disturbances to vegetation, wildlife, and habitat, human safety issues associated with AML are a major concern. There are obvious dangers associated with open pits and collapsing shafts and tunnels. There are less obvious hazards associated with abandoned mine workings, including headframes, equipment, poor quality or toxic air, and hazardous materials. In 2004 there were 34 fatalities nationwide associated with abandoned mine lands, nearly all from drowning and ATV crashes. Although the last death in New Mexico associated with an AML was in 2001 (when a high school student fell to his death in a 200-foot deep shaft near Orogrande), such fatalities and injuries occur regularly.

Water Quality

This is primarily an issue of contaminated runoff from mine sites, but it can be difficult to differentiate natural drainage conditions from those caused by mining. What we considered poor mining practices today, such as dumping mine wastes into drainages, were common in the past. Depending on the contaminants involved, their concentration and contact with living organisms, contaminated water has the potential to harm aquatic organisms as well as other plants and animals. Deposits that most typically cause drainage quality problems are base and precious metals, uranium, and high-sulfur coals, although not all such deposits produce water quality problems. Many others, including some industrial mineral deposits, can impact water quality. The major potential impacts of AML on water quality include:

- Acid drainage
- Metal leaching and resulting contamination
- Release of processing chemicals
- Increased erosion and sedimentation

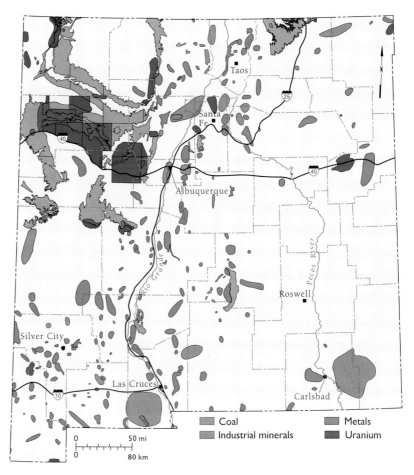

Mining districts in New Mexico. Although many of these districts are no longer active, this map provides a general idea of where abandoned mines are likely to exist.

The New Mexico Environment Department has identified twenty impaired streams in New Mexico (including the Red River) potentially affected by mining. Various federal and state agencies plan to remediate these areas, where necessary, to eliminate the impact of AML on the affected watersheds.

Societal Effects

Often there are competing societal issues involved with AML sites. The urgent necessity to remediate hazardous AML sites must sometimes be balanced with the historical significance of some sites and their importance to regional tourism. Mineral collecting in southern New Mexico, for instance, is dependent upon access to inactive mines and provides much needed income to communities including Deming, Lordsburg, and Silver City. Some towns are particularly proud of their mining history and do not always seek remediation of their AML sites. The residents of Leadville, Colorado, for instance, insisted on having mine waste and tailings piles remain. Special covers were designed that prevent adverse water quality impacts but maintain the characteristic look of historic mining. The same was true for residents of Madrid, New Mexico. Some historic mines are open for tours. The Harding mine in Taos County, long inactive, is currently maintained as a field laboratory and mineral collecting site.

AML PROGRAMS IN PLACE

There are a number of existing programs throughout the U.S. that address the remediation of AML. They are administered by federal, tribal, state, and local government agencies and private industry. The purpose of such programs is to inventory historic, inactive, and abandoned mines, identify and prioritize hazards associated with these mines, and remediate those hazards.

The primary source of funding for remediation of AML by governmental agencies is through the Surface Mining Control and Reclamation Act (SMCRA), signed into law on August 3, 1977. The act established a coordinated effort between the states and the federal government to fund abandoned coal mine remediation. SMCRA provided funding, through a tax on current coal production, to reclaim land and water resources adversely affected by pre-1977 coal mining. The act also allows funds to be used for remediation of non-coal mines. SMCRA only provides funding for states (like New Mexico) that produce coal. Other sources of funding must be obtained for remediation in non-coal producing states.

SMCRA established successful AML programs in coal-producing states to achieve these purposes. In addition, federal agencies, Native American tribes, and the mining industry also have successfully remediated many AML sites. There exists a proven record of successful mine land reclamation, hazard abatement, and effective management of appropriated AML program funds.

AML programs in place in New Mexico include:

The New Mexico Abandoned Mine Land Program

The New Mexico Abandoned Mine Land Program is administered by the Mining and Minerals Division (MMD) of New Mexico's Energy, Minerals and Natural Resources Department. This is the state program that receives SMCRA funding, usually $1.5–2 million each year. Established in 1980, this program has remediated over 4,000 of the most hazardous coal and non-coal AML features in the state. By law, human safety issues are at the top of the priority list, although environmental problems often are addressed in tandem with safety concerns. For example, backfilling shafts and pits with waste piles from the mine eliminates both features as well as their safety risk. In other cases, underground workings are left open where bats, owls, and other wildlife have taken up residence, but grates are installed over the openings that allow access by wildlife but not humans.

Although SMCRA allows funds to be used on non-coal sites, abandoned coal mines remain the highest priority. New Mexico still has an estimated $10-12 million worth of work to do to remediate its coal sites. These projects include hazardous mine openings, subsidence into old mine workings, and coal mine waste stabilization and reclamation in watershed areas. These cost estimates do not include the safeguarding and/or extinguishing of underground coal mine fires currently burning in the Gallup, Madrid, and Raton coal field areas. The exact location, extent of the area burning, potential loss to the coal resource, and cost of extinguishing these mine fires have not been determined but will be significant.

The program has received federal recognition three times in the past six years for its achievements in reclamation. Most recently, the AML program at MMD received the Best in the Western Region Award for 2004, for the Cerrillos South Mine Safeguard Project, located 25 miles south of Santa Fe. The innovative abandoned mine safeguarding measures in place here were part of the development of the first park in New Mexico dedicated to mining history, the Cerrillos Hills Historic Park. High-strength steel mesh covers with viewing platforms were installed over several shafts to allow visitors to safely view essentially untouched mine workings. Puebloan Indians were mining turquoise and lead in this area as early as 900 A.D., Spanish mining of lead and silver began in 1598, and for several years in the early 1880s there was an Anglo mining boom producing lead, silver, copper, and manganese.

The Navajo AML Reclamation Program

Native American tribes as well as states may establish AML programs and receive funding under SMCRA. From 1988 to 1992 the Navajo Abandoned Mine Lands Reclamation Program (NAMLRP) initiated an on-the-ground inventory assessment on non-coal-related AML sites. The Navajo Nation has jurisdiction

This grate over an abandoned mine shaft near Cerrillos allows bats access to the underground workings but prevents human entry. This feature is part of an award-winning AML project accomplished by the Mining and Minerals Division.

on Tribal Trust Lands only.

In 1989 Navajo AML initiated reclamation work on Priority 1 non-coal sites. Since then Navajo AML has successfully completed approximately 90 percent of the 1,085 inventoried non-coal AML sites. These non-coal sites include uranium, copper, and sand and gravel mines. The mine features include both surface mines such as open pits, rimstrips, and trenches, and underground mines such as portals/adits, and incline and vertical shafts. The terrain and environmental conditions varied widely from the low and dry lands of Cameron, Arizona, to the mountainous, rough, and wetter lands of the Chuska Mountains. Navajo AML has received five Office of Surface Mining awards for its reclamation efforts and numerous partnering opportunities.

After remediation of AML sites, SMCRA allows funding to be used for public works projects. Navajo AML initiated the Public Facility Projects (PFP) Program in 2000. In fiscal year 2002 they funded twenty PFP projects at approximately $4.8 million. This will ultimately account for $16.2 million in completed projects.

Bureau of Land Management, U.S. Forest Service, and National Park Service

The Bureau of Land Management (BLM), U.S. Forest Service (USFS), and National Park Service are responsible for managing most federally owned land, including remediation of AML sites. Each agency is developing inventories of AML sites and, as federal funding becomes available, these agencies have remediated AML sites on their lands. The USFS anticipates that funding will increase in 2006 for completion of inventory and continued remediation of sites on USFS land. The BLM estimates that 3,000 sites on public lands in New Mexico require remediation. These federal agencies work closely with the state AML program.

Other programs not funded through SMCRA include:

U.S. Army Corp of Engineers RAMS Program

The U.S. Army Corp of Engineers initiated the Restoration of Abandoned Mine Sites (RAMS) in 1999 to assist other agencies and industry in remediating AML sites. RAMS funded the New Mexico Bureau of Geology and Mineral Resources to complete an inventory of mines in Sierra and Otero Counties; this report will be released in 2005. RAMS also funded a database of remediation technologies. In addition RAMS is partnering with other agencies in developing plans for remediation of sites in the Red River and Pinos Altos

Remediation of abandoned sites involves surveying the extent of mine workings and associated hazards, as in this mine shaft near Orogrande.

areas. Funding for RAMS is federal, but separate from SMCRA.

Industry Programs

Many mining companies have remediated AML sites. Under the 1993 Mining Act any mine in New Mexico that had 24 months of production since 1970 is required to provide a plan for and implement reclamation and remediation. Many mining companies, including Phelps Dodge, Quivira Mining Company, Homestake Mining Company, St. Cloud Mining Company, and Molycorp have remediated historic mines on their permitted areas and in some cases other AML near their mines. Most active mining companies in the state are required to have an approved mine closeout plan, which provides for reclamation to a beneficial use after mining ceases.

REMAINING ISSUES FACING NEW MEXICO

Abandoned Mine Inventory

Most of the western states lack comprehensive inventories of abandoned mine sites, especially non-coal mines. In the early years of the AML programs, abandoned coal mine areas were inventoried extensively as a requirement of the program and in satisfying the development of approved reclamation plans and the prioritization of AML projects. Historical information regarding location, production, and ownership is generally more available for coal mines than it is for non-coal or hard rock mines.

CHAPTER TWO

The New Mexico Mines Database

The New Mexico Bureau of Geology and Mineral Resources, a service and research division of New Mexico Tech, serves as the state's geological survey. Since 1927 the bureau has collected published and unpublished data on the districts, mines, pits, quarries, deposits, occurrences, and mills in New Mexico. The bureau is converting that historical data into the New Mexico Mines Database, to provide computerized data that will aid in identifying and evaluating resource potential, resource development and management, production, and possible environmental concerns, such as physical hazards, indoor radon, regional exposure to radiation from the mines, and sources of possible contamination in areas of known mineral deposits. These data will be useful to federal, state, and local government agencies, public organizations, private industry, and individual citizens in order to make land-use decisions. These data are particularly useful in identifying mine sites in a given area and examining the potential for that mine site in contributing metals and/or other contaminants to the watershed.

The database provides information on the mines, quarries, mineral deposits, occurrences, mills, smelters, mine rock piles, tailings, and pit lakes located in New Mexico. Altered and mineralized areas are included because these areas have particular importance in terms of mineral resource development and/or environmental impacts. The database will be linked to other information such as geochemistry of samples collected by the bureau (and others).

The data have been gathered from published and unpublished reports and miscellaneous files in the bureau's mining archive. Information on location, production, reserves, resource potential, significant deposits, geology, well data, historical and recent photographs, mining methods, maps, and ownership are included. The database can be incorporated with other GIS layers, including geologic maps, topography, geophysical data, remote sensing data, the New Mexico Geochron database, and the New Mexico Petroleum database. Eventually the database will be accessible via the bureau's Web page; until then the database is partially available as bureau open-file reports. Although there are many other (and broader) applications for the information contained in this complex database, certainly the cooperative development of an inventory of AML sites in the state is one very important application.

Before New Mexico can fully address its AML issues, especially as they relate to physical hazards, it should have a comprehensive inventory of all abandoned mine sites. This would include locating, classifying, prioritizing, and incorporating mine features into a long-term plan for safeguarding and reclamation. Then a meaningful needs assessment can be compiled and the appropriate reclamation budgets developed. This inventory should be a multi-agency effort, using information already gathered by a number of state, federal, and tribal entities.

Funding

SMCRA fee collections drive the larger AML programs and are scheduled to sunset on June 30, 2005. Congress has not yet passed legislation that would establish a new fee schedule or formally extend the fee collection period. Congress did pass a continuing resolution, which extended the established fee collection temporarily. Without sufficient and predictable funding for the AML programs, many government agencies cannot adequately address hazardous mine features or reclaim and return AML to beneficial use. In addition, western states without coal mines do not receive SMCRA funding.

Baseline Conditions

One of the most difficult tasks in remediating historical mine sites is determining and characterizing the baseline conditions or natural background that existed before mining. A knowledge of baseline conditions is necessary, particularly in complex geologic settings, in order to establish meaningful remediation goals. Methods that may be used for this evaluation include integration of historical information, published values of unmined, mineralized areas, analyses of water from monitoring wells, leaching studies, statistical analyses, isotopic studies, identification of background by subtracting mining influences, and computer modeling. The U.S. Geological Survey has been the primary agency to apply science to identify background conditions. The New Mexico Environment Department contracted with the U.S. Geological Survey to characterize the baseline conditions along the Red River in Taos County to better understand complex interactions between naturally occurring geologic features and mining-related water impacts. The paper in this volume by Kirk Nordstrom outlines much of the work that was accomplished in that study.

Watershed Protection and Restoration in New Mexico—With a Focus on the Red River Watershed

David W. Hogge and Michael W. Coleman, *Surface Water Quality Bureau*
New Mexico Environment Department

The federal Clean Water Act was adopted in 1972 with the objective of restoring and maintaining the chemical, physical, and biological integrity of the nation's waters. The Clean Water Act is the basis of most national and state surface water quality standards and regulations. Like the federal act, the New Mexico Legislature provided objectives and policy direction for the protection of water quality when it adopted the state's Water Quality Act in 1967. Monitoring, assessing, and restoring water quality are key to successful implementation of both the federal and state acts, and these responsibilities are the foundation of the work of the Surface Water Quality Bureau of the New Mexico Environment Department.

The term *watershed* refers to the region that is drained by a given stream, lake, or other body of water. It describes the area that contributes water to a given stream or other water body. Any concern over water quality—or the impairment of water quality—in a given body of water must necessarily take into account the health of the entire watershed. The terms *catchment area* and *drainage basin* are often used interchangeably.

DETERMINING THAT A WATERSHED IS IMPAIRED

In accordance with the New Mexico Water Quality Act, the Surface Water Quality Bureau implements a comprehensive water quality monitoring strategy for surface waters of New Mexico. The monitoring strategy establishes methods to identify and prioritize water quality data needs, specifies procedures for acquiring and managing water quality data, and describes how these data are used to progress toward three basic monitoring objectives: to develop water quality based controls, to evaluate the effectiveness of such controls, and to conduct water quality assessments.

As in most other states, the Surface Water Quality Bureau uses a rotating basin system approach to water quality monitoring. Using this approach, selected watersheds are intensively monitored each year. The goal is to monitor every watershed in the state at least once every eight years. Revisions to the schedule may be necessary based on staff and monetary resources, which fluctuate annually. The Environment Department's monitoring efforts are also supplemented with other data collection efforts, such as USGS water quality gaging stations, which can be used to document long-term data trends.

Data collected during intensive surveys are used to determine whether state surface water quality standards are met and to ensure that designated uses are supported. Assessed data are used to develop the state's list of impaired waters (which is part of the *Integrated CWA §303(d)/305(b) Water Quality Monitoring and Assessment Report*) and Total Maximum Daily Loads (TMDL).

What Is a TMDL and How Is It Developed?

A TMDL, or Total Maximum Daily Load, sets an "allowable budget" for potential pollutants by scientifically determining through rigorous study the amount of pollutants that can be assimilated without causing a water body to exceed the *water quality standards* set to protect its designated uses (e.g., fishery, irrigation, etc.). Once this capacity is determined, sources of the pollutants are considered. TMDLs include both point and nonpoint sources. Once all sources are accounted for, the pollutants are then allocated, or budgeted, among the sources in a manner that will describe the limit (the total maximum load) that can be discharged into the river without causing the stream standard, or budget, to be exceeded. Nonpoint sources are grouped into a "load allocation" (LA) and point sources are grouped into a "wasteload allocation" (WLA). By federal regulation, the budget must also include a "margin of safety." Thus, 100 percent of the budget cannot be allocated to pollutant sources. The margin of safety accounts for uncertainty in the loading calculation. The margin of safety may not be the same for different water bodies due to differences in the availability and strength of data used in the calculations.

Water quality impacts come in many forms. Impacts can be from point sources of pollution—i.e., discharge that flows into a receiving body from a pipe or some other discrete source. Point source discharges include effluent from wastewater treatment plants, industrial discharges, and storm water associated with construction and industrial activities. Point sources are generally addressed through imposition of National Pollutant Discharge Elimination System permit limits, pretreatment requirements, management of storm water, and other discharge management strategies. Impacts can also be caused by nonpoint sources of pollution. Nonpoint source pollution, according to the U.S. Environmental Protection Agency, "occurs when water runs over land or through the ground, picks up pollutants, and deposits them in surface waters or introduces them into ground water." Sources are often indistinct, such as abandoned mines, agricultural runoff, erosion from denuded hillsides or streambanks, fire scars, overgrazing or overcutting, parking lots, recreational or paved roads, etc. Nonpoint sources of pollution are generally addressed through "best management practices" and other watershed restoration activities. Current estimates indicate that nonpoint sources are the cause of 93 percent of the state's surface water quality problems.

WATERSHED RESTORATION

Watershed restoration activities that address nonpoint sources of water pollution are generally non-regulatory, voluntary initiatives that are driven by a local desire to restore watershed health. Successful watershed restoration efforts rely, for the most part, on the strength of collaborative efforts to build a watershed community among local residents, agencies, and other stakeholders.

According to New Mexico's 2004–2005 Integrated Clean Water Act Report (section 303d and section 305b), probable causes and probable sources of watershed impairments include on-site liquid waste disposal, roads, recreation, urban storm water runoff, agriculture, ranching, silviculture, and resource extraction. Although no "hard" data exist, wildlife grazing (particularly by elk) is known to also contribute to localized water quality problems in certain areas of the state. Grazing and habitat alteration are the predominant sources of lake water quality impairment.

The implementation of treatment activities that reduce water quality impairment has been an effective tool in addressing watershed impairment. Treatments and controls for nonpoint source pollution are called *best management practices* or BMPs. BMPs can include constructed means of reducing impairments to surface and ground waters, such as inducing a more stable stream channel morphology with structures to deflect flows or installing a sewer system to replace individual septic systems in a community. Nonstructural BMPs are conservation practices related to the way in which we manage our resources. The timing and rate of fertilizer and pesticide application, instituting storm water management ordinances, or creating a rotation system for cattle grazing in areas where ground cover is critical for preventing soil erosion are examples of these. BMPs should realistically represent the best combination of structural or nonstructural management practices working together to reduce impairments to water quality. BMPs should be based on conditions of the site where the practices are to be constructed and/or implemented and should be based on the economics and performance targets associated with the specific problem to be addressed.

RED RIVER WATERSHED RESTORATION EFFORTS

The Red River Watershed is a major tributary to the upper Rio Grande in northern New Mexico. Twenty-one perennial watercourses, draining an area of 226 square miles, originate as very high quality mountain streams. The Carson National Forest manages approximately 90 percent of the area. Elevations range from 13,161 feet at Wheeler Peak (the highest point in New Mexico) to 6,500 feet at the confluence with the Rio Grande. The lowest four-mile reach of the Red River flows through a spectacular canyon of the Wild and Scenic River Area that includes the Rio Grande gorge. The only towns within the watershed are Questa and Red River, which at an elevation of 8,750 feet is the highest incorporated town in the state.

The Red River has long been recognized by state and federal agencies as a high priority watershed. It occupies one of the most popular multiple use watersheds in the state, devoted to recreational activities—chiefly skiing and fishing—along with widespread livestock grazing by U.S. Forest Service permittees. Legacy mining and exploration, as well as development, extraction, and processing of world-class mineral deposits, are other prime features of the watershed. Concerns include: mining (primarily Molycorp, and to a lesser extent legacy mining sites in tributary drainages); septic tank leach fields in the alluvial valley above the town of Red River; unlined sewage lagoons in the village of Questa; leaking underground petroleum storage tanks in the town of Red River; and sediment contributed by steep, bare slopes at the Red River ski area and from many dirt roads, grazing allotments, and hydrothermal scar areas on the national forest.

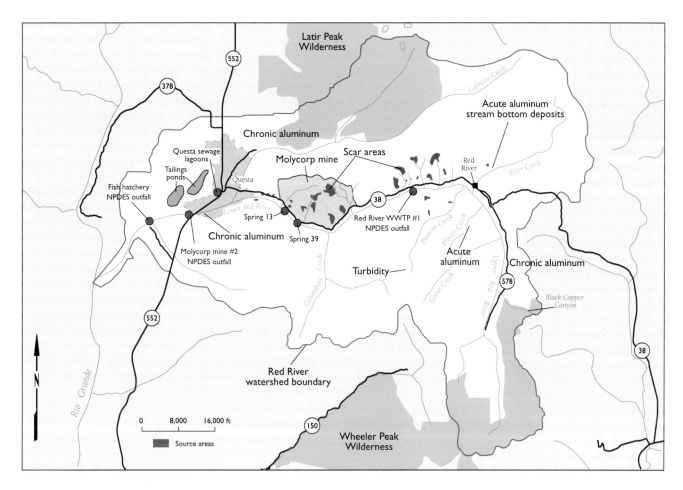

Map of Red River Watershed, Taos County, New Mexico. Impaired reaches and the pollutants of concern are shown.

RED RIVER WATERSHED PROJECTS

Using Clean Water Act Section 319 grant funding, the Surface Water Quality Bureau embarked upon a series of on-the-ground implementation projects to address impacts to the watershed. These projects began in 1991, and several are underway at the present time.

Beginning in 1995 the Surface Water Quality Bureau helped to initiate the formation of a Red River Watershed Association. Meetings were initially held in Red River and Questa, with attendance and participation by interested citizens, state and federal agencies staff, environmental groups, and municipal representatives. Following reorganization in 1998 the Red River Watershed Group has continued to draw together a broad-based group of watershed residents, agencies, and stakeholders to take on the immense task of restoring conditions that will improve the quality of water—and therefore the quality of life—throughout the watershed. The group addresses a variety of water quality issues throughout the entire drainage—from the headwaters to the Rio Grande—through a collaborative, consensus-based approach in which every voice has equal weight.

The Red River Watershed Group's major focus is:

- To determine pollutants, their sources and effects, and communicate the information to citizens

- To seek opportunities to enhance fish habitat within the watershed

- To bring citizens together to restore, protect, and fully utilize the Red River

- To educate and inform users and citizens about the area and watershed stewardship

Restoration projects in the Red River Watershed have been funded primarily through federal Clean Water Act Section 319 dollars, along with a tremendous effort by local volunteers and local, state, and federal agencies. Projects initiated in the Red River Watershed include:

Mineral Extraction Impacts

This project was initiated in the early 1990s and was one of the Surface Water Quality Bureau's early efforts to use passive limestone anoxic drains to intercept and treat acid mine runoff. This first drain was installed at the Oro Fino mine, along upper Bitter Creek.

Red River Ground Water Investigation

This project identified and addressed, via BMP implementation, several forms of nonpoint source pollution impacting the Red River.

The Red River in the vicinity of Molycorp mine is a gaining stream system, recharged throughout the length of its main stem by shallow ground water. Seeps and springs entering surface water were determined to be virtual point sources of contamination, posing a sizable impairment to the Red River. The primary solution implemented was to install an anoxic alkaline drain along the seep areas, which proved to be effective in neutralizing the acidic, heavy metal-bearing seeps before they could mix with the surface flows of the Red River.

Lower Bitter Creek Restoration Project

This interagency cooperative pollution prevention project was developed on lower Bitter Creek, a perennial-intermittent tributary to the Red River. BMPs to control erosion in the stream channel, along local roads, across unstable slopes, and at the toe of an active hydrothermal scar's landslide zone, were designed by the project cooperators. U.S. Forest Service crews, contractors, Youth Conservation Corps participants, and volunteers completed the on-the-ground work, which resulted in a measured decrease in turbidity, sediment loading, and heavy metal delivery at the Bitter Creek–Red River confluence.

Enhanced Local Involvement for Addressing Water Quality in the Red River Watershed

This ongoing effort to gain broader and more effective local participation from throughout the Red River watershed addresses significant water quality issues, with the goals of developing a cost-effective watershed cleanup strategy, and identifying and prioritizing sites for cleanup. Composed of key stakeholders, the Red River Watershed Group's work will address impacts identified through the TMDL process. Strategies developed will provide a framework for addressing and reducing pollutant loading from both public and private lands.

River Park Stream Rehabilitation Project, Town of Red River

This project is designed to address heavy sediment loads that have caused the active Red River channel to expand horizontally and become very shallow. A heavily impacted 1,500-foot section of the river is being restored to a more functional width/depth ratio and sinuosity using rock flow-management structures, willow plantings, and other BMPs. The project will increase the river's scour energy, enabling it to transport its sediment load, while improving the long-term stability of the channel and bank.

Collaborative Red River Restoration Off Road Vehicle Impact Remediation

This will reduce sediment and turbidity caused by unrestricted off road vehicle use by: implementing and maintaining a series of BMPs: obliterating and reclaiming temporary or unauthorized roads, controlling surface erosion at recreation sites, managing off road vehicle use and enforcing recreation regulations, increasing public outreach and education on water quality protection at recreation areas, and revegetating disturbed areas.

Suggested Reading

Coleman, M.W., 2000, Lower Bitter Creek Restoration Project: Summary report for FY 94-B grant project; submitted to U.S.E.P.A., Region 6, Dallas, in completion of Clean Water Act section 319(h) project; New Mexico Environment Department, Surface Water Quality Bureau, Santa Fe.

Hopkins, S., 2003, Special water quality survey of the Red River and tributaries, submitted to U.S.E.P.A., Region 6, Dallas, in completion of Clean Water Act section 106 grant; New Mexico Environment Department, Surface Water Quality Bureau, Santa Fe.

Slifer, D., 1996, Red River ground water investigation: final report for FY 92-A grant project; submitted to U.S.E.P.A., Region 6, Dallas, in completion of CWA section 319(h) project; New Mexico Environment Department, Surface Water Quality Bureau, Santa Fe.

State of New Mexico, 2004–2006, Integrated Clean Water Act §303(d)/§305(b) Water Quality Monitoring and Assessment Report; New Mexico Environment Department, Surface Water Quality Bureau, Santa Fe.

Acid Rock Drainage

Kathleen S. Smith, *U.S. Geological Survey*

Acidic drainage is a common water quality problem associated with hard rock metal mining and coal mining. Acidic drainage (commonly called "acid rock drainage" or ARD) is formed when rocks containing sulfide minerals come into contact with water and oxygen to create sulfuric acid, which in turn releases metals (e.g., aluminum, manganese, copper, zinc, cadmium, arsenic, and lead) from the rocks. This acidic metal-laden water can negatively impact water supplies used by municipalities, agriculture, or wildlife. Understanding the formation of acidic drainage involves the fields of geology, chemistry, and biology. However, the fundamental source of acid and metals in the drainage is rocks, including the host rocks of a mineral deposit and waste rocks resulting from mining.

There are several phrases used to describe water affected by the weathering (wearing away or erosion and chemical decomposition) of rocks in mining and mineralized areas. The term "acid rock drainage" covers both mining related and naturally formed acidic drainage, and is used in this report to emphasize that not all drainage affected by the weathering of rocks is related to mining. The term "acid mine drainage" (AMD) is limited to drainage that is both acidic and mining related. The term "mining influenced waters" (MIW) is limited to drainage that is mining related, but not necessarily acidic. This term is useful because not all drainage from mining areas is acidic, but non-acidic drainage may still contain significant concentrations of metals.

HOW DO YOU MEASURE ACID?

The pH scale is a measure of the amount of acid, with acids having pH values less than 7 and bases having pH values greater than 7. The pH scale is logarithmic, so water with a pH value of 3 is ten times more acidic than water with a pH of 4, and one hundred times more acidic than water with a pH of 5. Most natural waters are in the pH range of 5 to 9. Rain has a pH of approximately 5.7 because carbon dioxide from the air dissolves in raindrops to form a weak acid. Many familiar liquids also are acidic. For example, lemon juice and vinegar are both acidic and have pH values of approximately 2 and 3, respectively. Mining-influenced waters can have a wide range of pH values and are not always acidic. Young fish and some aquatic insects may be harmed by pH levels below 5. The pH also can affect aquatic life indirectly by changing other characteristics of water chemistry.

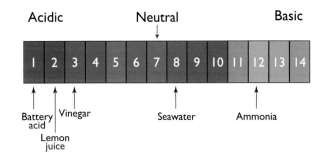

The pH scale, a measure of acidity, showing the pH of some common liquids.

WHAT DOES MINING HAVE TO DO WITH ACID ROCK DRAINAGE?

One of the factors that has the greatest influence on the production of ARD is rock type. Rocks are made of minerals, naturally occurring chemical compounds that have specific crystal structures and documented chemical compositions. Different minerals have different properties and weather differently in the presence of water. For example, some minerals, such as halite (table salt), readily dissolve in water, whereas other minerals, such as quartz (found in beach sand), are practically inert (not subject to change). Various metals are contained within the structure of these minerals, and release of these metals into the environment depends on the properties of their resident minerals.

A mineral deposit is an accumulation of metals or minerals in a relatively small volume of the earth's crust. Mineral deposits form as a result of large-scale flow of metal-bearing fluids deep in the earth's crust. The processes involved in depositing the minerals are collectively called mineralization. When mineralized rock is mined, it exposes new surfaces that can be

Photo of the Carlisle mine in the Steeple Rock district of southwestern New Mexico. Geological characterization (below) can be very helpful in predicting potential environmental effects of mined sites.

GEOLOGICAL CHARACTERISTIC	TYPE OF INFORMATION ACQUIRED
Geologic setting	pH buffering capacity Ease of subsurface transport Routes to biological receptors
Mineral deposit type	Which metals are present Acid-generating capacity
Historical mine/mill activities	Predict contaminants of concern (e.g., mercury, cyanide) Efficiency of sulfide removal from wastes

Some aspects of geological characterization. Buffering minimizes pH changes in water when acid or base is added.

weathered and accelerates the production of ARD.

Mineral deposits can be categorized into different types, defined by characteristic minerals and associated potential environmental impacts. Some kinds of mineral deposits tend to produce very acidic, metal-laden waters, whereas others tend to produce less acidic waters with fewer or different dissolved metals. The particular metals and minerals present in rocks and waste rock are characteristic of how that mineral deposit was formed, and of the regional rock type, hydrology (how solutions move through the rocks), and geologic structures (such as faults). Geologic characterization can be very useful in predicting the potential environmental impact and footprint (impacted area) of a mined site.

Several characteristics of water that are common in drainage from mining and naturally mineralized areas include: low pH (acidic), elevated sulfate concentrations, elevated iron, aluminum, and/or manganese concentrations, elevated concentrations of other metals, and high turbidity, which is a measurement of the amount of small particles suspended in water. These components commonly depend on the mineralogy of the deposit.

WHAT KINDS OF ROCKS MAKE ACID?

Sulfide minerals are common in many types of hard rock metal mining and coal deposits. Many sulfide minerals are relatively unstable under surface conditions, so when they are exposed to air and water they undergo the chemical reactions of weathering. Pyrite, or fool's gold (iron sulfide), is a common sulfide mineral that produces acid when it weathers. The generation of ARD begins with a startup reaction. For example, when pyrite comes into contact with water and oxygen (in air) the result is a reaction that produces dissolved iron and sulfuric acid (a mixture of sulfate and acid). In the acidic drainage generation cycle, once the startup reaction has begun, acid production is self-propagating as long as pyrite, water, and microorganisms (bacteria) are present. The acidic drainage generation cycle involves converting iron from one form (iron II) to another form (iron III). This conversion has been called the rate-determining step (the bottleneck) because it is slow at low pH. However, certain kinds of microorganisms can greatly speed up this conversion of iron by as much as 100 to 1,000,000 times, especially at low pH. Once the ARD reactions have begun, conditions are favorable for microorganisms to speed up the conversion of iron and lessen or eliminate the bottleneck in the acidic drainage generation cycle. This is important because iron III can readily react with pyrite (and some other sulfide minerals) and produce more acid. In fact, weathering via iron III can produce eight times more acid than weathering via oxygen (as in the startup reaction). So, the faster the iron can be converted from iron II to iron III, the more acid can be produced.

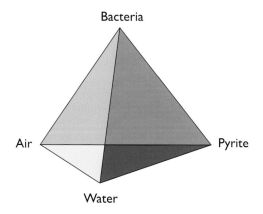

The Acid Rock Drainage (ARD) tetrahedron, showing the relationship between the four components that produce ARD. All four components are required to produce ARD.

IS ACID THE ONLY PROBLEM?

The acid produced from pyrite weathering can attack other minerals in surrounding rocks. For example, acidic water can attack and dissolve common rock-forming minerals (e.g., mica and feldspar) and produce dissolved aluminum. Manganese and calcium also are common elements that have elevated concentrations in ARD.

In mined and mineralized areas it is common to see solid precipitates (residues) of various colors coating stream bottoms. These precipitates can result in high turbidity, which makes the water cloudy and interferes with sunlight penetration into the water (which in turn interferes with photosynthesis by aquatic plants). Iron forms coatings on stream bottoms over a broad range of pH conditions, and the coatings can vary in color from yellow, to orange, to deep red; these iron coatings are called *yellow boy*. The different colors are different iron-bearing minerals that form under different pH and chemical conditions; these minerals can be used as indicators of the chemical conditions present in the stream. Aluminum forms white coatings on stream bottoms above a pH of around 5. Precipitated aluminum may harm fish by accumulating on their gills. Manganese forms dark brown or black coatings at higher pHs, usually above 7. These various precipitates can form by natural processes in unmined areas, and their colors in stream bottoms were used as a prospecting tool by early miners to identify mineralized areas. Stream-bottom coatings may damage the habitat, inhibit growth, or kill aquatic organisms that live on the bottoms of streams.

WHERE DO METALS COME FROM IN ACID ROCK DRAINAGE?

Trace elements (e.g., copper, lead, zinc, cadmium, arsenic, selenium) are normally present in low concentrations in the earth's crust. However, in mineralized areas certain elements (depending on the deposit type) are present in above-average concentrations. Many of the trace metals in mineralized areas are found in various sulfide minerals. Minerals that contain trace metals can be weathered or attacked by acidic water or iron, thereby releasing metals into the environment. The concentration of a released metal is a function of (1) the concentration of that metal in the mineral, (2) the accessibility and susceptibility to weathering of the minerals that contain the metal, and (3) how easily the metal can be transported through the environment under the existing conditions.

The confluence of Deer Creek with the Upper Snake River near Montezuma, Colorado. As low-pH water containing dissolved aluminum mixes with higher-pH water, the aluminum precipitates to form a white coating on the streambed.

Metals differ from organic contaminants in that they do not break down in nature. Therefore, once metals are released into the environment, they persist. However, metals may be affected by physical and chemical processes that can modify their transport through the environment. Once metals are released from their parent mineral, they can be transported by water or sediment, precipitated, or taken up by solid particles. Generally, metals are more easily transported in lower-pH waters (pH less than 4 or 5), which is why acidic drainage usually contains elevated concentrations of metals. However, some metals (e.g., zinc and cadmium) can be transported in near-neutral pH waters (pH 6 or 7), and some elements (e.g., arsenic, molybdenum, and selenium) can be transported under higher-pH conditions (pH greater than 7 or 8).

In mineralized areas and mining-related rocks, trace metals and acid may be temporarily stored in salts formed from evaporating mineralized solutions. Once wet conditions return, these salts can readily dissolve to release acid and metals. Dissolved metals also may be incorporated into solid particles; later changes in chemical conditions, such as pH, may cause the particles to release these metals back into the water.

Diagram showing how the concentration of base metals (zinc, copper, cadmium, lead, cobalt, and nickel) vary with pH in natural and mine waters. Diagram courtesy of Geoffrey Plumlee and Sharon Diehl, U.S. Geological Survey.

HOW DO YOU REMEDIATE ACIDIC WATERS?

Most mineral weathering reactions use up acid instead of making acid. Some minerals dissolve better than others or use up more acid when they dissolve. The mineral calcite (calcium carbonate), which is the mineral that makes up limestone, is very effective at neutralizing (counterbalancing) acidic solutions. This is because it dissolves fairly easily in acidic solutions and uses up acid when it dissolves. So, calcite commonly is used in the treatment of acidic waters. Moreover, it is a common mineral in some rocks and can neutralize acidic waters as they flow through those rocks. Other minerals also can help neutralize acidic solutions, but they tend to be less effective than calcite because they dissolve more slowly.

The final pH of the drainage from mining and mineralized areas is a balance between acid-generating reactions (e.g., pyrite weathering) and acid-neutralizing reactions (e.g., calcite weathering). The relative rates of these reactions and the accessibility of the reacting minerals ultimately determines the pH and metal content of the drainage water.

WHAT TYPES OF MINING CAN PRODUCE ACID ROCK DRAINAGE?

Mineral deposits and the ores within them come in a variety of sizes and shapes. Therefore, methods used to mine a particular mineral deposit must be tailored to the size, shape, depth, and grade (richness) of the ore being mined. There are two main types of mining: surface mining and underground mining. In both, as minerals are removed, rock surfaces are exposed to water and air. This presents the opportunity for acid-producing pyrite weathering reactions to proceed and produce ARD. Therefore, openings from underground mines, such as adits, tunnels, and shafts, may drain ARD. Warning signs of ARD near mines include red-orange-yellow coatings on rocks, dead vegetation, and green slime growing in discharge waters. Even though not all discharges from mined areas are acidic, they may still contain significant concentrations of dissolved metals.

Compared to other industries, mining is unique in that it usually discards more than 90 percent of the material that is processed. Therefore, there is a lot of solid waste rock associated with mining, and the characteristics of this rock control its potential environmental impacts. Mining-related rocks from both underground and surface mining may produce ARD. Once fresh rock surfaces are exposed to water and air, the minerals in the rocks can weather. Historic mines generally did not have the technology to efficiently remove or isolate acid-producing minerals or metal-rich minerals from the mining-related rock. In addition, mine rock piles were commonly put in the most convenient place for the miners, generally close to the mining operation, which could be in or adjacent to a stream or drainage, or at the angle of repose on a mountainside. Therefore, many ARD water quality problems are associated with older mines.

WHAT IS OPEN-PIT MINING, AND HOW DOES IT LEAD TO PIT LAKES?

Open-pit mining is a surface-mining technique that is used when the orebody is large and relatively near the surface. Open-pit mining involves repeated removal of layers of rock (both ore and overburden) to form a large open bowl-type structure. Several New Mexico copper and gold mines were mined by open-pit methods. Many open-pit mines exceed 1,000 feet in depth; therefore, most of the large open-pit mines extend well below the ground water table. Once mining and dewatering have ceased, these open-pit mines commonly fill with water to form pit lakes. At some point, pit lakes generally reach the point where the amount of inflow water approximately equals the amount of outflow water.

Pit lakes can receive inflow from both surface and ground water. Ground water models predict most pit lakes to be terminal basins, which means that they pull in water from all sides and evaporatively concentrate potential contaminants in the lake. During this evaporation process, water is evaporated and contaminants that were in the water remain behind; so, contaminant concentrations increase because there is less water present to dilute them. If this is the case, and there is no outflow from the pit lake (i.e., if it acts like a sump), then potential contaminants from the open-pit mine are contained within the lake. However, if there is outflow from the lake, potential contaminants may flow down gradient from the lake. Once a pit lake is filled, it may persist and fluctuate in elevation with seasonal changes in weather and water flow.

The New Mexico Mines Database includes a table listing thirteen current pit-lake areas in New Mexico. Some pit lakes have water quality that is suitable for recreation, such as the Copper Flat pit lake near Hillsboro, New Mexico. Other pit lakes have acidic waters with elevated metal concentrations, such as the Chino pit lake near Silver City, New Mexico. Many technical questions and issues remain about accurate prediction of the hydrology and water quality of future pit lakes.

Suggested Reading

Brady, K. B. C., Smith, M. W., and Schueck, J., eds., 1998, Coal mine drainage prediction and pollution prevention in Pennsylvania: The Pennsylvania Department of Environmental Protection. Available online www.dep.state.pa.us/dep/deputate/minres/districts/cmdp/main.htm

Hem, J. D., 1989, Study and interpretation of the chemical characteristics of natural water, 3rd edition: U.S. Geological Survey Water-Supply Paper 2254, 263 p.

Hudson, T. L., Fox, F. D., and Plumlee, G. S., 1999, Metal mining and the environment, AGI Environmental Awareness Series 3: Alexandria, Virginia, American Geological Institute, 64 p.

Plumlee, G. S., and Nash, J. T., 1995, Geoenvironmental models of minerals deposits; fundamentals and applications, in du Bray, E.A., ed., Preliminary compilation of descriptive geoenvironmental mineral deposit models: U.S. Geological Survey Open-File Report 95-0831.

Schmiermund, R. L., and Drozd, M. A., eds., 1997, Acid mine drainage and other mining-influenced waters (MIW), in Marcus, J.J., ed., Mining environmental handbook: London, Imperial College Press.

A River on the Edge—Water Quality in the Red River and the USGS Background Study

D. Kirk Nordstrom, *U.S. Geological Survey*

Mineral resource production is vital to modern, industrialized societies. Unfortunately, mining and mineral processing can cause serious damage to water, air, soil, and biological resources. Acid mine waters, produced from mines that extract valuable metals such as copper, gold, silver, zinc, and lead, have damaged aquatic life, crops, and livestock. These problems are not irreversible; lands and waters disturbed by mining can be restored, but often at considerable cost.

An important challenge related to mine-site reclamation and cost-benefit analysis is the "natural background" or pre-mining water quality. The pre-mining conditions can provide a justifiable objective for cleanup goals. However, pre-mining water quality rarely was measured at mine sites before the 1970s, and even if it had been measured, the methods of sample collection, preservation, and analysis were likely much less reliable than current methods. Before 1970 detection limits alone would have been higher than most current water quality standards for metals. Consequently, any attempt to determine retroactively the natural background water quality at a mine site today depends on indirect methods using scientific inference.

MINE CLOSURE REGULATIONS AND THE USGS BACKGROUND STUDY

The New Mexico Mining Act of 1993 and the New Mexico Water Quality Act (1967) require operating mines to meet several regulations on closure. One requirement is that ground water must meet New Mexico ground water quality standards *unless* it can be shown that ground water before mining contained solute concentrations greater than the standards. For such a site, the natural background values, rather than the standards, may be used.

One of the largest and most productive molybdenum mines in the U.S. is operated by Molycorp, Inc., near Questa. To provide technical information needed to help settle disputes between regulatory agencies and Molycorp regarding pre-mining ground water quality at this mine site, the U.S. Geological Survey (USGS) conducted a study in cooperation with the New Mexico Environment Department. The project has taken more than three years (2001–2005) and involved an interdisciplinary team of experts in economic and environmental geology, mineralogy, geochemistry, hydrology, geomorphology, and geophysics. It is in the first stage of completion. To the best of our knowledge, this study is the first to estimate pre-mining ground water quality for regulatory purposes at an active mine site by a third party.

THE KEY: A PROXIMAL ANALOG SITE

The approach taken by the USGS was to study a proximal analog site in detail and apply the knowledge gained to the mine site. A proximal analog is located off the mine site but nearby with the same geologic, hydrologic, and climatic conditions as the mine site, and whose water-rock interactions would provide a viable model for the mine site. Any substantial differences between the analog site and the mine site are identified and accounted for with appropriate models of water-rock interactions. The Straight Creek Basin was chosen for the analog site (Figure 1).

During the USGS background study, there was extensive sampling of ground waters and surface waters, especially a detailed chemical survey of the Red River. As the ultimate recipient of ground water flow in the Red River Valley, the Red River water chemistry contains clues on where ground water enters the river, and its composition.

WATER QUALITY OF THE RED RIVER

More than one million years ago, the Red River was the headwaters of the Rio Grande. The Red River begins at an elevation of 12,000 feet near Wheeler Peak, the highest peak in New Mexico, and flows 35 miles to the Rio Grande. A USGS gaging station is located at the U.S. Forest Service Ranger Station at Questa with hydrograph records that date from 1924 to the present. Daily discharges average 46.8 cubic feet per second with a large range of 2.5–750 cubic feet per second. Areas of highly altered rock (known as "scars") erode so rapidly that vegetation cannot be sustained. Acid waters occur naturally in scar areas from

ENVIRONMENTAL AND WATER QUALITY ISSUES

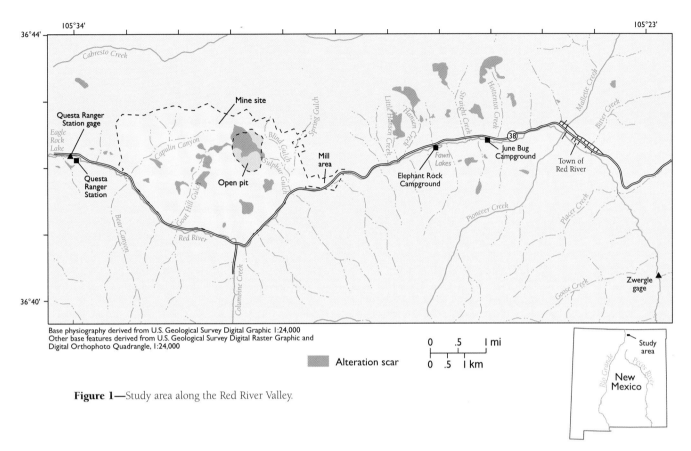

Figure 1—Study area along the Red River Valley.

the weathering of pyrite (iron sulfide).

During August 17–24, 2001, the USGS sampled water along the Red River as part of a constant-injection tracer study to determine the discharge and solute-concentration profile with distance. All major ions and several trace elements were determined on filtered and unfiltered samples including iron, aluminum, manganese, copper, zinc, cadmium, lead, arsenic, cobalt, nickel, molybdenum, and beryllium. These elements are of particular concern because they can be discharged from mining activities and can cause harm to aquatic biota in the river. The analytical results, the most detailed chemical survey ever done for the Red River, are available online at

http://wwwbrr.cr.usgs.gov/projects/GWC_chemtherm/pubs/OFR03_148_QuestaTracer.pdf.

One indication of the aquatic health of the Red River is shown in Figures 2 A, B, and C. Profiles of discharge, pH, and alkalinity with distance downstream from the town of Red River are shown in Figure 2A. Note that the river flow tends to increase down drainage, not continuously but in steps at specific points in the river. The big step increase at about 13,000 meters is the point at which Columbine Creek enters the Red River. The smaller increase at 6,000 meters, however, has no obvious tributary entering the river and must be from ground water inflows. Note that just upstream from the 6,000-meter point the discharge is decreasing; it is a "losing" stream where the river water is being lost to the subsurface. This loss is caused by stream flow entering the large debris fans that push out from their respective drainages into the Red River and cover part of the river. Water flowing through the debris fans emerges farther downstream. The fans cause a "damming" effect with sediments depositing behind them. Sediments deposited behind the Hottentot fan formed the flat valley for the town of Red River.

The pH measurements provide an estimate of the acidity or basicity of the water. Values of pH less than 7 are acidic and greater than 7 are basic or alkaline. A pH near the neutral point of 7, say 6–8, is healthy for aquatic life and for human health. The pH values for the Red River tend to be in the 7.5–8.5 range and indicate good water quality. Note, however, the distinct decreases, or dips, in pH especially at the three points marked by the vertical dashed lines in Figure 2. These are points where acid ground waters enter. The first pH dip at Waldo Springs is where naturally occurring acid ground waters from the Hansen, Straight, and Hottentot scar drainages enter the Red River. The second dip is near the Sulphur Gulch

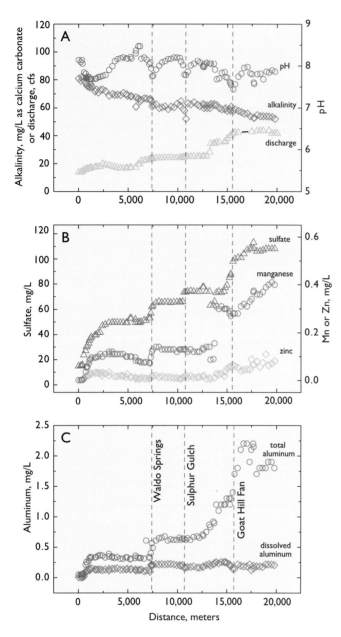

Figure 2— A Discharge, pH, and alkalinity with downstream distance in the Red River from the town of Red River to the USGS gaging station. **B** Sulfate, manganese, and zinc concentrations with downstream distance in the Red River from the town of Red River to the USGS gaging station. **C** Total (dissolved plus particulate) and dissolved aluminum in the Red River from the town of Red River to the USGS gaging station.

drainage where acid waters enter from the mine site (part natural and part related to mining), and the third dip is where acid waters enter from the Goat Hill Gulch area (likely natural). Both natural scar drainage and mine-waste-pile leachates occur at the mine site. However, major seeps are now being intercepted and are no longer entering the Red River.

Alkalinity is another measure of the health of the Red River. Alkalinity is an estimate of the buffering capacity (or neutralizing potential) of the water; the higher the alkalinity, the greater the capacity of the water to resist changes in pH from acid inflows. The profile in Figure 2A shows a steadily decreasing alkalinity with downstream distance. Clearly the addition of acid inflows along the Red River is using up its buffer capacity, making it more susceptible to acidification.

Figure 2B shows distance profiles for concentrations of dissolved sulfate, manganese, and zinc with notable step increases at Waldo Springs, Sulphur Gulch, and Goat Hill Gulch. Sulfate is derived from dissolution of the mineral gypsum (calcium sulfate), which is common in the Red River Valley, and from weathering of pyrite. Only pyrite weathering, however, produces acidic waters, which can occur both naturally and from mine wastes. Hence, pyrite weathering has caused some of the increases in sulfate at these step increases because they coincide with the same places where the pH decreases occur. Manganese and zinc also have step increases in the same places as the sulfate increases, because the minerals from which these elements weather (rhodochrosite and sphalerite) are minerals that accompany pyrite and gypsum. Both of these minerals occur in the Questa ore deposit, rhodochrosite in substantial amounts. Concentrations of most other dissolved metals in the Red River are too low to be of concern for aquatic health standards.

Figure 2C is a distance profile for dissolved and total (dissolved plus particulate) aluminum. Aluminum, derived from common rock-forming minerals, is highly soluble in acid waters but becomes insoluble when the pH increases to about 5. Acid inflows to the Red River are neutralized upon mixing, and a white aluminum precipitate can be seen in many places along the banks. Consequently, the total aluminum concentration in the river builds up as suspended particles, but the dissolved aluminum remains at a low and nearly constant concentration because of the insolubility of this hydrous aluminum precipitate at neutral pH values.

WHEN GOOD RIVER QUALITY GOES BAD

Although the quality of the Red River is affected by acid inflows, most of the time a healthy pH and adequate alkalinity are maintained. Rapid, deleterious changes in the water quality occur when summer monsoonal rainstorms hit the valley. An example is shown in Figure 3 for a rainstorm event of September 2002. During the early part of the rainstorm, the pH decreased from 7.8 to 4.8, sulfate concentration doubled, and manganese concentration increased fourfold. Another example was recorded in 1986 that resulted in even greater increases in acidity and metal concentrations. These sudden changes in water quality are caused by a surge of acid drainage that usually comes from natural scar areas upstream from the mine site, but which may have come from leaching of mine wastes before remedial action was taken. Large quantities of suspended sediment also are released during these high-flow rainstorm events, which can have a deleterious impact on aquatic life. Fortunately, these are short-term changes, on the order of a few hours or less, and the river does recover. For this short duration, an increase in acid inflow overcomes the buffering capacity of the river, and that is why the water quality is marginal or "on the edge." Although water quality in the river generally is adequate for aquatic life, the river has little resistance to perturbations, such as rainstorms or additions of large quantities of mine waste effluent.

CONCLUSIONS FROM THE USGS BACKGROUND STUDY

There is no question that in the highly mineralized Red River Valley, acidic ground waters from scar weathering occur naturally with high concentrations of metals, sulfate, and fluoride that can be at least ten

Aluminum precipitation along the north bank of the Red River near the Molycorp mine site.

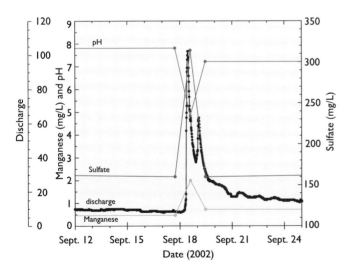

Figure 3—Change in discharge, pH, sulfate, and manganese at the USGS gage during the storm of September 17, 2002.

times greater than ground water quality standards. There is also no question that mine-waste effluent has contributed additional acidity, metals, sulfate, and fluoride to local ground waters at the mine site. The USGS study of the Straight Creek analog site, combined with other data collected in the Red River Valley, is providing information that constrains pre-mining ground water concentrations and can provide a baseline or reference for setting site-specific standards. These pre-mining ground water solute concentrations vary somewhat from catchment to catchment, and even within different parts of the same catchment depending on the geological characteristics. Hence, a single concentration value (per constituent) for the whole mine site probably is not a feasible approach in such a complex and heterogeneous terrain. A range of concentrations per constituent and per catchment is necessary to characterize the mine site ground waters and to allow for uncertainties in the estimate. Although these conclusions create a more cumbersome regulatory process, they reflect the realities of the complex geology and hydrology in this environment.

Water Quality Regulation of Mining Operations in New Mexico

Mary Ann Menetrey, *New Mexico Environment Department*

The New Mexico Environment Department is the primary state agency regulating water quality protection at hard rock mining sites in New Mexico. Over the last several decades there has been increased awareness of the environmental impacts associated with mining operations, including contamination of ground water and surface water resources. Protection of this water is critical, given New Mexico's limited surface water resources and heavy reliance on ground water for public water supplies. Ninety percent of New Mexico's population depends on ground water aquifers for drinking water, and nearly 80 percent of the population is served by public systems derived from ground water sources. Surface water flows in many New Mexico rivers have been significantly reduced due to drought conditions and increased water demands, and New Mexico will continue to rely more heavily on ground water to sustain the state's residential population and business community.

The importance of the state's water resources makes the implementation and enforcement of water quality regulations a top priority. Mining operations are commonly situated in complex and sensitive environments, making water quality protection a challenge. Below is a summary of the water quality regulations that affect hard rock mining operations, as well as a discussion of how the regulations apply to contaminant sources, application of water quality standards, and mine closure.

WATER QUALITY LAWS AND REGULATIONS AFFECTING MINING OPERATIONS

The first water quality protection law in New Mexico, the Water Quality Act, was adopted by the New Mexico legislature in 1967. The Water Quality Act was amended in 1973 to allow the state of New Mexico to adopt regulations requiring that permits be obtained for water quality protection. These regulations went into effect under the jurisdiction of the Water Quality Control Commission (WQCC) in 1977 and provide the current framework for New Mexico's water quality protection programs.

The Water Quality Control Commission regulations include numerical standards for ground water and surface water, and are designed to protect all ground water in New Mexico that has a total dissolved solids concentration of 10,000 milligrams per liter (mg/L) or less. The regulations provide for water quality protection using two methods. The first method is through discharge permits that prevent exceedances of numerical water quality standards by controlling sources of contamination. The second method is through abatement plans that address cleanup of existing contamination. The Environment Department's Ground Water Quality Bureau is responsible for administration of the WQCC ground water regulations as they apply to discharges from mining operations and other types of facilities.

The ground water discharge permit provisions of the WQCC regulations are the foundation of New Mexico's ground water pollution prevention programs. These provisions require that a person discharging onto or below the surface of the ground demonstrate that the discharge will not cause ground water standards to be exceeded at any place of withdrawal for present or foreseeable future use, and will not cause any stream standard to be violated. At mine facilities the regulated discharges can include process solutions, waste rock, mill tailings, leach stockpiles, storm water, and domestic wastewater. The Ground Water Quality Bureau currently has discharge permits for approximately fifty mine facilities, including facilities for mining and/or processing of uranium, copper, molybdenum, gold, and other metal-bearing ores. The primary components of each discharge permit are an operational plan, monitoring plan, contingency plan, and closure plan. The goal of the permitting process is to work cooperatively with operators to keep ground water contaminants contained, and to ensure leaks and spills are detected early and promptly remediated.

The state of New Mexico also works closely with the U.S. Environmental Protection Agency (EPA) in implementing the federal Clean Water Act, Safe Drinking Water Act, and other federal laws that address water quality protection. In particular, the EPA administers the National Pollutant Discharge Elimination System (NPDES) permit program that is applicable to mining

ENVIRONMENTAL AND WATER QUALITY ISSUES

operations pursuant to the Clean Water Act. These NPDES permits are intended primarily to protect surface water quality and address point source discharges to surface waters. The Environment Department's Surface Water Quality Bureau coordinates with EPA in administering the NPDES program by certifying permits, conducting inspections, and providing permit information to the public and operators. The Surface Water Quality Bureau is in the process of obtaining primacy for the NPDES program, which will allow the state to issue NPDES permits to mining and other facilities. Nonpoint source contamination at mine sites is addressed through implementation of best management practices and storm water controls.

SOURCES OF GROUND WATER AND SURFACE WATER CONTAMINATION

Ground water and surface water contamination have occurred at many mining operations throughout the state. Although the sources of this contamination vary, some of the primary contributors include acid rock drainage from sulfide-bearing ore and waste rock, as well as process solutions that have escaped from unlined leach stockpiles and tailing impoundments. Much of the existing contamination from mining facilities is a result of past disposal practices that would not be permitted under the current regulations. Recently issued permits have focused on more rigor-

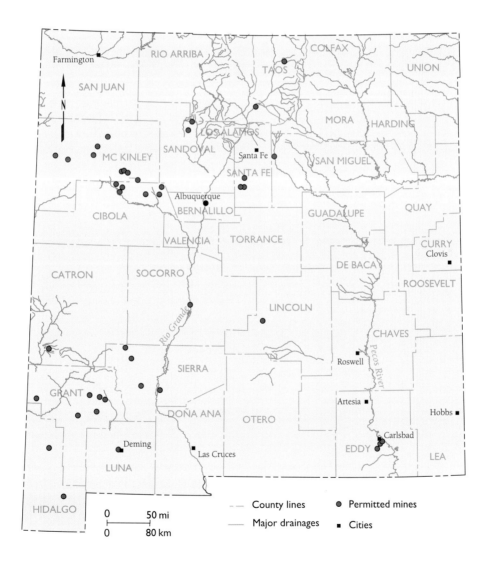

Mines with ground water discharge permits in New Mexico.

ous contamination prevention measures, such as lining of leach stockpiles and establishing waste rock management plans to reduce the potential for acid rock drainage.

Metals are the primary contaminants at mine sites affected by acid rock drainage. These metal contaminants include aluminum, copper, cadmium, arsenic, chromium, fluoride, and zinc. This contamination is a result of acidic solutions releasing metals contained within stockpiled rocks or tailings exposed to water and oxygen. Acidic solutions are intentionally applied to ore piles at some of New Mexico's copper mines to speed up this acid rock drainage process that dissolves copper so that it can be removed for processing. Total dissolved solids and sulfate are also common contaminants, often found as a precursor to metal contaminant plumes or associated with mine wastes that lack the potential for acid generation.

Open pits associated with mining operations can also be a source of water quality degradation. To facilitate mining, many large open pits have been mined below the ground water table, and dewatering operations are necessary to keep the pits dry. Where the walls of these pits contain sulfide-bearing minerals, oxidation of these minerals causes the release of metal contaminants into surface runoff waters that can accumulate in the pit bottoms. When mining ceases and dewatering of the pits stops, ground water flow back into the pits can create pit lakes that exceed surface water standards for many contaminants.

APPLYING WATER QUALITY STANDARDS AND ABATEMENT

Once ground water or surface water becomes contaminated, cleanup can be very challenging and costly. Many mine facilities are located over fractured bedrock, where it is difficult to install extraction wells to recover the contaminated water. Contaminated waters move through these fractures and can eventually migrate into aquifers that provide a current or future water supply, or they can enter surface waters. Additionally, tailing impoundments and leach stockpiles often contain very large volumes of residual process solutions that can take decades or longer to completely drain from the piles. Where these facilities are unlined and water has become contaminated, it is almost impossible to prevent this drainage from contaminating underlying ground water for years to come.

The Water Quality Control Commission has determined that most ground water in New Mexico with a total dissolved solids or TDS concentration of less

Pit lake at Nacimiento open-pit copper mine near Cuba.

than 10,000 mg/L, including the ground water directly underlying mine facilities, has a reasonable and foreseeable future use. Therefore, the ground water underlying tailings, leach stockpiles, and other mine source areas must be cleaned up to water quality standards if it becomes contaminated. Contaminated water beneath these facilities cannot be left unabated because ground water can migrate away from these facilities and contaminate nearby water supplies. Also, many mine sites are candidates for future industrial or residential development, and in either case will need a clean on-site water supply.

The Water Quality Control Commission recognized that there might be situations where it is not technically or economically feasible to fully abate contamination of ground water. The WQCC regulations address these situations by including provisions for operators to petition the WQCC for approval of alternative abatement standards (AAS), which are a type of variance from the numerical ground water quality standards. In order to obtain alternative abatement standards the petitioner must demonstrate that:

- compliance with the applicable WQCC ground water standards is not feasible or there is no reasonable relationship between the economic and social costs and benefits;

- the proposed AAS are technically achievable and cost-benefit justifiable; and

- compliance with the proposed AAS will not create a present or future hazard to public health or undue damage to property.

The Water Quality Control Commission has approved alternative abatement standards for two mine sites in New Mexico, including the L-Bar uranium mill site in Cibola County and the Cunningham Hill mine in Santa Fe County. The Environment Department was able to support the AAS petitions for both these sites because, in part, the mine operators had conducted extensive ground water abatement, characterization, and source control measures before submitting their petitions. Given the difficulties of cleaning up existing contamination at many other mine sites, the Environment Department anticipates that there will be several more AAS petitions coming before the WQCC in upcoming years.

Another important consideration regarding ground water cleanup at mine sites is the issue of background concentrations. Many mine operations are situated in mineralized areas where ground water may be naturally elevated in concentrations of total dissolved solids, sulfate, or certain metals. Under the abatement provisions of the WQCC regulations, where the background concentration of any contaminant exceeds the numerical ground water standard, the responsible person can abate to the background concentration rather than numerical standards. However, determining the background concentration can be extremely complicated at mine sites where the geology is complex and there are multiple aquifers. Due to site-specific differences, the Environment Department does not believe it is appropriate to establish a single method for determining background concentrations. Background concentration investigations are ongoing at several mine facilities, including the Molycorp Questa mine, where the U.S. Geological Survey has been conducting a background investigation since 2001.

WATER QUALITY PROTECTION FOLLOWING MINE CLOSURE

One of the greater challenges facing state regulators and mine operators is determining adequate closure methods for existing operations after mining opera-

Seepage into the Red River at Spring 39 adjacent to the Molycorp mine site.

tions cease. In addition to the long periods needed for drainage of process solutions, acid rock drainage at some facilities is predicted to continue for decades and possibly centuries. Under the WQCC regulations, ground water discharge permits must include closure plans to ensure that water quality standards are met after the mine operation shuts down. The components of the closure plan are a description of closure measures, maintenance and monitoring plans, post-closure maintenance and monitoring plans, financial assurance, and other measures necessary to prevent and abate contamination.

The Environment Department has consistently required that closure plans for mining operations focus on enacting source control measures to reduce or eliminate ongoing contamination of ground water and surface water after operations cease. Containment of contamination without source controls would not meet WQCC requirements and is not practical for sites where contamination could continue for decades or centuries. Measures that are typically included in mine closure plans approved by the Environment Department include long-term stabilization measures such as regrading of stockpiles and erosion controls, covers over stockpiles and tailings that minimize infiltration into contaminated material, and a plan for long-term water treatment of contaminated ground water and surface water.

Due to the large size of some mining operations and extent of existing contamination, there is uncertainty as to whether approved closure measures will successfully protect water resources; therefore, many approved closure plans also include requirements for additional studies to refine the closure plan and allow the mine to plan better for closure. Examples of additional studies include test plots, stability studies, hydrologic investigations, and feasibility studies. Approximately thirty studies related to site closure are currently underway for the Molycorp Questa mine. As more data are collected in the area of mine closure, it is anticipated that permitting for closure under the Water Quality Control Commission regulations should become both more effective and more efficient.

Engineering Challenges Related to Mining and Reclamation

Terence Foreback, *Mining and Minerals Division*
Energy, Minerals and Natural Resources Department

Mining in New Mexico has been going on for hundreds of years. During this time the public's perception of the concepts of environment, water, economics, and sustainable development has changed. A hundred years ago, mines were not planned, operated, or reclaimed according to today's standards. More recent operations that were planned or operated without sustainable development in mind now are challenged by regulatory requirements to reclaim to a stable landform. With the passage of mining legislation such as the New Mexico Mining Act in 1993 and the Surface Mining Control and Reclamation Act of 1977 (SMCRA), new mining operations in New Mexico are required to integrate planning, operating, and reclaiming activities with a specific post-mine land use in mind. For this reason, the major engineering challenges must be viewed in the context of the age of the operations, and reclamation standards that have changed over time. The New Mexico Mining Act distinguishes between "existing units" and "new units" when applying reclamation standards. Some engineering challenges are unique to older sites. Newer sites can be planned from the beginning to accomplish sustainability of the post-mine land use. Some of the major engineering challenges include:

- Slope stability and reclaiming mined land to landscapes that are geomorphologically stable and self-sustaining

- Design of operations and reclamation strategies to minimize negative public perception issues

- Reclaiming mine sites to geochemically stabilize land without impacting water quality

SLOPE STABILITY

Landforms in New Mexico consist of hills and valleys, mesas and arroyos, mountains and canyons, which are all ways of referring to low areas and high areas. Slopes at various angles connect the low and high areas to each other. Slopes are never completely stable because of the effects of gravity and erosion, which are constantly trying to create a flat environment. We've

Successfully reclaimed mined land in New Mexico during the second growing season.

all had some experience with slope stability working in our yards or in a sand pile, and we can understand that we can stack some materials higher and steeper than other materials. The steepest angle at which a pile of material will stand is referred to as the "angle of repose." All materials have an internal resistance or friction. The higher the internal resistance, the better the material's ability to resist the effects of gravity. Materials with higher internal resistance will have a steeper angle of repose, and different materials have different angles of repose. For any material, though, the lower the slope angle, or said another way, the flatter the slope, the more stable the slope.

When we discuss slope stability at mining operations, we are generally referring to the reclaiming of mined landscapes to be geomorphologically stable. Geomorphologically stable means creating a stable topography that soil, slope, and weather would naturally form over time, similar to the natural landform. Mining regulations now require that mine reclamation create a self-sustaining ecosystem that includes the physical stability of the landscape. This has not always been the case. As a result, some older mines have been abandoned and left with unstable slopes. Some active mines must now reclaim unstable slopes that were created before mining regulations required such reclamation.

So what types of slopes are we talking about reclaiming at mining operations in New Mexico?

- Slopes on spoils (in-pit waste rock) at coal mines
- Slopes on out-of-pit waste rock piles at open-pit metal mines (i.e., Tyrone)
- Slopes on tailings dams
- Any slope that was created or affected by the mining operation that must be reclaimed

Slopes on Spoils at Coal Mines

Reclamation of spoil at coal mines involves restoring natural vegetation and drainage in order to return the mine site to a self-sustaining ecosystem such as farmland, wildlife areas, parklands, or housing developments. The reclamation process actually begins before the first ton of coal is removed. To fulfill mining permit requirements, the coal company must document how sedimentation from the temporarily disturbed areas will be controlled, how ground and surface waters will be protected, and how restoration of the soil and vegetation will be achieved.

Fluvial geomorphic-based design of post-mining topography is an engineering technique that is being used at mining operations in New Mexico. The end product of fluvial geomorphic-based design is a post-mining topography that resembles the natural surroundings. The reclamation grading and recontouring goals of fluvial geomorphic-based design are:

- To provide long-term stability for steep slopes and drainages
- To increase topographic diversity to improve habitat for plants and wildlife
- To reduce the potential for flash flooding of adjacent property as compared to the pre-mine conditions
- To create a functional landform that blends in with the surrounding natural terrain

The fluvial geomorphic approach creates a stable topography similar to what the combination of soil, slope, and weather would naturally form over time and includes sinuous drainage systems that mimic surrounding terrain. Channel dimensions are designed to pass the storm discharges that would be expected from a yearly maximum storm event and to pass the

Overview of Cottonwood pit reclaimed topography in the foreground and undisturbed topography in the distance. Reclamation work is shown prior to revegetation.

sediment from these annual flow events. These channels are also designed to pass larger events without excessive erosion by allowing overbank discharges to spread out over a flood-prone area as would naturally occur.

San Juan Coal Company, a New Mexico mining operation, received the 2004 federal Office of Surface Mining's National Award for Excellence in Surface Coal Mining, Best of the Best Award for its application of fluvial geomorphic-based techniques. Although this procedure has been pioneered at New Mexico's coal mines, it can be applied to other types of mines where conditions allow.

Slopes on Out-of-Pit Waste Rock Piles at Open-Pit Metal Mines

When open-pit metal mines remove non-economic material overlying an orebody, the material must be placed somewhere out of the way of the mining operation. This out-of-pit material is referred to as waste rock. Waste rock piles can vary in size depending on the mining operation, but some waste rock piles can be quite large.

Waste rock piles in the past were placed in the most convenient location determined solely on economics. They were typically dumped at their angle of repose on pre-existing hillsides near the mining operation. Sometimes these hillsides were not stable. In some locations waste rock was deposited on material that trapped ground water flowing through the waste rock piles or contained clay and developed into a sliding surface. These factors have all contributed to unstable waste rock piles that now must be addressed. Mitigation of the waste rock piles that were not designed to be geomorphologically stable is difficult. Mitigation is a process of lessening (reducing) the slope of the pile, combined with ensuring adequate drainage.

There are two ways to reduce the slope of a waste rock pile. Earthmoving equipment can start at the top and extend the toe of the stockpile in length, thus creating a flatter surface. If this is not possible because of constraints such as highways, topography, structures, streams, or environmental concerns, then the material must be removed starting at the top and placed in another location.

Ensuring adequate drainage lessens the effects of surface and sub-surface water. Constructing surface interception ditches can reduce infiltration by preventing surface water from flowing onto rock piles and by preventing ponding of water on rock piles. Mitigation of ground water effects is difficult on rock piles that were not properly engineered.

Waste rock piles at new operations can be designed to be geomorphologically stable. Slopes can be constructed at an angle much less than the angle of repose. Rock pile locations can be selected so that material is not placed on areas that are already unstable. Drainage under rock piles can be established by constructing gravel underdrains that can prevent the buildup of ground water pressures. Surface drainage can be designed to prevent any ponding of water on rock piles, and diversions can carry water away from the pile.

Many mining operations can also be designed to place some or all of the waste material back into the open pit. The advantages of partial backfill are possible reduction of disturbed area, containment of waste rock piles, and partial reclamation of the open pit. The backfilled material can then be graded as described above in order to achieve stability.

Slopes on Tailings Dams

Tailings dams are man-made structures used to contain the non-ore material that is left after ore is processed. Typically, this waste material is very fine and has a high moisture content.

The same types of considerations discussed above apply to slopes on tailings dams. The major engineering design difference is that tailings dams are structures containing large volumes of material consisting of water and very fine material created during the processing of the ore. The slopes on the tailings dams must be not only geomorphologically stable, but the dam must hold back material that has low internal strength. For this reason, the state of New Mexico has regulations and standards on the design and construction of these structures.

Other Slopes at Mining Operations

There are other types of slopes at mining operations such as slopes on leach stockpiles. Current regulations require that these slopes be designed to be geomorphologically stable upon reclamation. The guiding

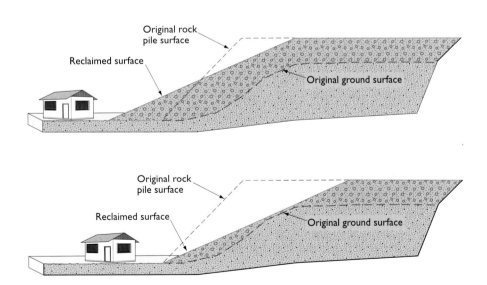

Grading waste rock piles to obtain more stable slope angles.

principles for the engineering design are creating flatter slopes and ensuring adequate drainage.

DESIGN TO MINIMIZE NEGATIVE PUBLIC PERCEPTION ISSUES

Public perception issues related to mining are often the concerns most frequently dealt with by regulators and legislators in New Mexico. These concerns are related to how the public comes into contact with mining issues such as:

- Dust from mining operations
- Blasting and its effects
- Traffic from mining operations, particularly from trucks
- Visual effects

Mining operations usually cannot eliminate these effects, but mines can be planned, operated, and reclaimed to minimize them.

Dust from mining operations can be controlled through a variety of engineering practices. Roads can be designed using materials that create less dust. There are environmentally friendly substances that can be applied to road surfaces to suppress dust. One of the simplest procedures is to apply water to the road surface. Mineral processing plants can use cyclones or other collection methods to capture dust rather than emitting it to the atmosphere.

Mining operations where blasting is performed can use blasting techniques that minimize its effects. Using blast delays that reduce the amount of explosives that are shot at one time can greatly reduce the effects of ground vibration. Reducing ground vibration can eliminate damage beyond the mining property boundary and minimize the perception of damage. Using blasting methods that reduce noise can greatly influence the public's perception of blasting. Proper planning, design, and operations can greatly reduce the public's perception of the consequences of blasting.

Traffic from mining operations almost always creates public perception issues. Usually the issue is trucks carrying the mined material, and the problem is particularly evident with aggregate (sand and gravel) operations. Solutions to this problem may be solicited in public meetings with affected stakeholders and could include minimizing dust from roads, carefully planning the hours of operation, monitoring the speed and size of the trucks, and minimizing (to the degree possible) the number of trips by processing on site.

Visual effects are often a concern to the public. But mines can be planned, operated, and reclaimed to minimize visual effects. Operations can be planned to minimize the amount of surface disturbance by mining only the required area to meet production requirements and by reclaiming areas expeditiously after mining. Planning should be given to the permanent mine facilities relative to their siting. Proper reclamation techniques previously discussed, such as fluvial geomorphic-based design, can also create a post-mine environment with a pleasing visual effect.

RECLAIMING MINE SITES TO GEOCHEMICALLY STABLE LAND WITHOUT IMPACTING WATER QUALITY

Another engineering challenge is reclaiming mine sites to be geochemically stable without impacting water quality, specifically:

- Soil problems related to coal mines
- Acid rock drainage from waste rock piles
- Soil problems related to reclaiming areas with low pH and high metal content.

To a large degree, if a mine is not reclaimed to be geochemically stable, it will not be geomorphologically stable.

Soil Problems Related to Coal Mines

Two major problems encountered during coal mine reclamation are salinity and clay content of topsoil and spoil material. Reclamation at some mines in New Mexico was problematic because vegetation could not be established in soils that contained a high salt and clay content. Regulations now require that mine operators sample the available topsoil material before mining. Both quality and quantity of the material must be addressed to ensure that both are adequate for reclamation. The material that will be mined must also be analyzed for problems in soil chemistry, which must be taken into account prior to reclamation.

Acid Rock Drainage from Waste Rock Piles

Waste rock piles may contain materials that create acid drainage. This usually is caused by water reacting with waste rock of material containing the mineral pyrite. The water flows through the rock pile, reacts with the pyrite, and then flows out of the rock pile as acidic ground water. Acidic water can dissolve metals and other substances and may contain unacceptably high levels of these contaminants. The acid rock drainage (ARD) can remain beneath the surface, causing contamination of ground water, or surface as springs and contaminate surface water.

The effects of ARD from rock piles can be minimized, particularly when mining operations plan, operate, and reclaim with mitigation in mind. Surface water can be diverted around rock piles to minimize the amount of water that will infiltrate into the pile. Soil layers placed on top of the rock piles can be properly designed and vegetated to use the precipitation that naturally falls on the rock piles without allowing infiltration. This is referred to as a "store and release cover." In extreme cases, systems can be designed and constructed to collect ARD that flows from rock piles so that it can be pumped and treated.

Soil Problems Related to Reclaiming Areas with Low pH and High Metals Content

Metal mining operations sometimes have unique engineering challenges reclaiming areas with low pH (high acidity) and high metals content. Two areas usually of concern are tailings areas and waste rock piles. There are several challenges with ensuring that proper engineering is done in these areas including:

- Limiting water infiltration and ARD by establishing a store and release cover.

- Ensuring that neutral cover material is placed over material that would inhibit plant growth. The concern here is that plants may not grow adequately in the acidic material or they may absorb metals through the roots.

- On areas that were not regulated and designed according to current standards, interceptor systems for ARD may need to be constructed after problems have been encountered.

REQUIRING ADEQUATE FINANCIAL ASSURANCE

Mining has left an unfortunate legacy of abandoned, unreclaimed sites in New Mexico. For this reason the New Mexico Mining Act and the Surface Mining Control and Reclamation Act require that mining companies calculate the cost of reclamation and post a bond for that amount in the event the company becomes insolvent and cannot meet its regulatory responsibilities.

The calculation of the cost to reclaim is an important engineering challenge. Engineers working on the calculation must have an adequate understanding of the principles of reclamation. There must be regulatory oversight of the calculation in order to ensure that adequate reclamation funding is available. This is one of the most important engineering challenges and one that ensures that unreclaimed sites will not be an issue of the future.

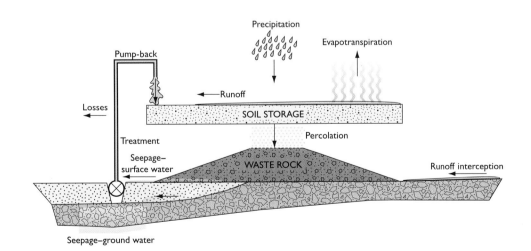

Conceptual design of an acid rock drainage collection system.

Environmental Impacts of Aggregate Production

William H. Langer, *U.S. Geological Survey*
L. Greer Price and James M. Barker, *New Mexico Bureau of Geology and Mineral Resources*

Our nation's infrastructure—streets, highways and bridges, houses and buildings, sidewalks, sewers, power plants, dams, just about everything—requires huge volumes of aggregate. The main sources of aggregate in the United States are sand and gravel (43 percent) and crushed stone (57 percent). These materials are commonly used with a binder in concrete, mortar, and blacktop. About 2.66 billion metric tons of natural aggregate worth $14.5 billion (exceeding the total value of all metals) were produced in the United States during 2003.

Many agencies advocate that resource extraction be sustainably managed. This approach commonly requires that decisions regarding resource extraction include best management practices and other environmental management tools that consider health, economic prosperity, and social well-being.

Most aggregate is used in its natural state, except perhaps for crushing, sizing, and washing. Unlike most metallic resources, aggregate is not concentrated from an ore. Quality aggregate is chosen specifically to avoid metallic and other minerals that react with water to create acid.

Because aggregate is expensive to transport, it is quarried near the point of use, commonly a population center. Resource availability and conflicting land use issues severely limit areas where aggregate can be developed. The siting of aggregate quarries depends primarily upon where the geology is favorable and not necessarily where it is needed most.

Most environmental impacts from aggregate quarrying are benign. The impacts are generally close to the quarry and can be largely mitigated. Impacts from most extraction activities relate to the geology of the site, and geologists can help identify impacts and design plans to avoid them. Aggregate production may alter geologic conditions, altering the dynamic equilibrium of the site. Specific impacts and their mitigation are outlined below.

CONVERSION OF LAND USE

The most obvious environmental impact of pit/quarry operations (which is what most aggregate "mines" consist of) is the conversion of land use, generally from undeveloped or agricultural lands. Surface pits and quarries are dramatically different from most other land uses. Careful selection of a pit or quarry site minimizes the amount of surface area that must be disturbed by resource extraction. In some cases, the post-quarry use is more acceptable to the public and more environmentally or economically valuable than the original, pre-quarry use.

Conversion of land use may impact features of special interest or significance. Conducting a pre-quarry inventory of the site for scenic, biological, historical, archaeological, paleontological, or geological features minimizes this impact. Ironically, such features are sometimes recognized as unique only when aggregate operation begins. Aggregate extraction uncovers a relatively large area at a relatively slow pace, which can lead to serendipitous discoveries to which many organizations are able to respond rapidly.

STATE	SAND & GRAVEL PRODUCTION (10^3 metric tons)	SAND & GRAVEL AS A % OF AGGREGATE PRODUCTION
Arizona	59,800	84.8
Colorado	33,500	73.0
New Mexico	13,900	77.0
Oklahoma	9,890	17.3
Texas	81,500	43.5
Utah	26,100	77.3
United States	1,140,000	42.9

Sand and gravel production for the year 2003, and sand and gravel as a percentage of the total aggregate production by states in the New Mexico region and for the United States.

CHANGE TO VISUAL SCENE

Conversion of land use often changes visual character, viewed either from the site or toward the site. The change, temporary or permanent, is subjective; what is acceptable to some is objectionable to others. Visual impact depends on the topography, ground cover, and the nature of the quarry operations.

Visual impacts can be mitigated through careful quarry planning and design. The process may include limiting extraction areas and unnecessary disturbance (such as road widths), staining fresh rock faces to resemble weathered rock, creating buffer zones, and visual screening such as berms, tree plantings, fencing, or using other landscaping techniques. Overburden and soil can be stockpiled out of view. Good housekeeping practices, such as maintaining equipment and locating equipment below the line of sight or in enclosed structures, are also effective.

LOSS OF HABITAT

Site preparation often results in loss of habitat for some species. Pre-production site inventories identify rare or endangered species so that habitat can be set aside, or so that extraction operations can be suspended during critical breeding or migrating seasons. Creation or improvement of habitat off site can offset the loss of habitat on site. Selected animals or plants can be relocated. After closure, the site may be reclaimed to look and function like the original habitat. Reclamation during quarrying, rather than waiting to the end of operations, can speed habitat recovery.

Many aggregate operations preserve existing habitat through creation of buffer areas. The buffer areas of many aggregate operations retain all the characteristics of the original habitat or may be planted to increase vegetative cover. In some populated areas, quarry buffers are a significant part of the total available open space. Wildlife from the surrounding area may seek the protection afforded by such buffers. Some active aggregate operations and their buffer zones can serve as habitat for rare or endangered species. Water is a major limiting factor in arid and semiarid climates. Irrigation may be necessary to establish new vegetation.

NOISE

The most frequent complaint from the public about aggregate operations is noise. Tolerance to new noise depends upon the background noise to which one has adjusted. In an urban or industrial environment, background noise may mask noise from an aggregate operation. In contrast, the same level of noise from an operation in a rural area or quiet residential neighborhood is noticeable to those accustomed to quiet settings.

Ambient noise generally is an accumulation that does not have a single, identifiable source. If noise can be identified as coming from a quarry, the perception of this noise may be enhanced. Noise impacts are highly dependent on the sound sources, the topography, land use, ground cover, and climate. Sound travels farther in cold, dense air and during atmospheric inversions.

The primary sources of noise during aggregate extraction are engines, processing equipment, and blasting. Aggregate producers are responsible for ensuring that the noise emitted from the pit or quarry does not exceed regulated levels. Regular inspections and maintenance can help ensure effective noise control for equipment.

Noise generated during quarrying can be mitigated through various engineering techniques. Topography, landscaping, berms, and stockpiles can form sound barriers. Noisy equipment (such as crushers) can be located away from populated areas and can be enclosed in sound-deadening structures. Conveyors can be used instead of trucks for in-pit movement of materials.

Trucking of aggregate is a significant source of noise. The proper location of access roads, the use of acceleration and deceleration lanes, use of engine-brake mufflers or avoidance of engine-brake use, and careful routing of trucks all can reduce this noise or at least its detection. Noisy operations can be scheduled or limited to certain times of day.

DUST

The impact of dust is determined by proximity of the operation to residential areas, ambient air quality, moisture, air currents and prevailing winds, the size of the operation, and interaction with other dust sources. Regulations strictly limit the amount of dust emitted during quarrying. A carefully prepared and implemented dust control plan reduces impacts. Controlling fugitive (non-point source) dust commonly depends on good housekeeping more than on elaborate engineered controls. Techniques of dust control include applying water and chemicals to haul roads, sweeping, reduced vehicle speed, windbreaks, and ground cover. Point source dust can be controlled using dry or wet suppression equipment. Dry suppression includes conveyor covers, vacuum collection systems, and bag houses. Wet suppression systems consist of pressurized surfactant-treated water sprays throughout the plant. Lack of an adequate water supply can present a problem for large operations in the arid west and thus water is often trucked or piped long distances.

BLASTING

Quarry blasting may occur daily or as infrequently as

Aerial view southward of the Taos Gravel Products gravel pit near the edge of the Rio Grande gorge. This operation was closed in 2002 because it was determined to be too close to the Rio Grande Wild and Scenic River Corridor and was situated (in part) on un-permitted land. The site has since been reclaimed.

once or twice a year. Potential impacts include ground vibrations, noise, dust, and flyrock. Geology, topography, and weather all affect the impacts of blasting. Buffer zones, tree belts, and berms may serve multiple purposes in reducing noise and dust levels between the mine site and community, while improving the visual quality of the area.

The modern technology of rock blasting is highly developed, and when blasting is properly conducted, the environmental impacts should be negligible. By following widely recognized and well-documented limits on ground motion and air concussion, direct impacts are mitigated.

CHEMICAL SPILLS

Routine equipment maintenance and blasting may result in the accidental spillage of solvents, fuels, and blasting agents, which can contaminate surface or ground water. Leaking underground storage tanks can pollute ground water. Minimizing chemicals used, properly storing all hazardous chemicals and petroleum products carefully within bermed areas, monitoring water for nitrates, and providing workers with training in safe operation and maintenance procedures are all part of best management practices.

GROUND WATER

Predicting the environmental impacts of these operations on ground water is highly dependent on an understanding of local geology, hydrology, and climate. Precipitation may flow into a quarry and recharge ground water. In dry climates, evaporation of water in pits or quarries may actually lower the water table. Removing vegetation from the quarry reduces evapotranspiration, which may ultimately increase ground water. In highly permeable deposits, impermeable subsurface (slurry) walls are sometimes necessary to isolate the pit from the water table. Water removed from pits through pumping can be returned to nearby streams, which may recharge the ground water supply downstream.

In some areas of aggregate production, changes in ground water quality have been attributed to the removal of soil that previously acted as a protective layer that filtered or otherwise reduced contaminants reaching the ground water. Many heavy metals, easily degraded organic substances, and bacteria are retained relatively well in the natural soil layer. If an underlying gravel layer is exposed, this retention is much weaker. The level of impact depends on the thickness and character of material removed, the surface area involved, and the total volume and recharge of the aquifer. Impacts can be mitigated by controlling water recharge in quarries or by locating quarries outside of recharge areas.

SURFACE WATER

Aggregate operations entail removal of vegetation that may retard runoff. Aggregate extraction may create impervious land that prevents infiltration or may change runoff patterns in other ways. Pits and quarries may affect surface water chemistry, but these subtle changes are primarily local. The ability to predict flooding and deposition at a pit or quarry largely depends on how well the hydrology and history of the adjacent stream and surrounding watershed are known.

Water from aggregate processing and storm runoff over pit/quarry sites can increase the suspended rock particles (turbidity) in stream runoff. Turbidity is generally greatest at pit/quarry and wash-plant water discharge points and decreases downstream. Turbidity can be controlled by filtering, or by containing runoff or wash water at recharge basins.

Aggregate production within stream floodplains may impact stream-channel morphology. Flooding streams may flow through a pit or quarry in an active floodplain resulting in permanent changes in channel position that cause bank erosion and undercutting. This can substantially alter the distribution of the energy and force of the stream. Levees or dikes protect pits/quarries from flooding and keep water and sediment in the main channel. Engineered spillways allow controlled flooding and prevent deposition of sediment in pits/quarries.

Few aggregate operations in New Mexico occur within active streams. Careful hydrologic studies and application of best management practices can allow aggregate to be extracted from active stream channels with little environmental impact. The type and severity of impacts are dependent on the geologic setting and characteristics of the stream. The main impact occurs if more sediment is removed than the stream can replenish. Sediment removal may be at one site or be the total of many smaller operations. Sediment removal changes the stream cross section and increases gradient at the pit, which may cause widespread upstream erosion and loss of riparian habitat. A decrease in stream sediment by deposition in a pit can cause the stream to erode, resulting in similar effects. After aggregate extraction ceases, stream recovery can be quite fast or take many years.

EROSION AND SEDIMENTATION

Quarrying can promote erosion, which can result in increased sediment in nearby streams. Slope stability, water quality, erosion, and sedimentation commonly are controlled by sound engineering and geologic decisions. Appropriate slope angles are important. Roads, drainage ditches, and operational areas must fit the particular site conditions. Disturbed areas can be protected with vegetation, mulch, or other cover. They can be protected from storm water runoff by the use of dikes, diversions, and drainage ways. Sediment can be retained on site using retention ponds and sediment traps. Regular inspections and maintenance help ensure continued erosion control.

LAND SURFACE

Aggregate operations should avoid areas of known landslides and areas favorable for mass movement. Aggregate operations on an existing landslide or near the toe or head of a landslide can remobilize the slide. In areas where natural factors are not conducive to slope failure, aggregate extraction can cause landslides if the pit or quarry is poorly located.

If a landslide does occur, it is likely to be near (but not necessarily at) the quarry. Landslides are likely to occur after quarrying starts, usually triggered by precipitation. This could be a single event or a series of landslides over an extended period of time. Geologic engineering techniques can identify existing landslides and landslide-prone areas, but they cannot predict precisely where or when a landslide will occur.

POST-QUARRYING IMPACTS

Most aggregate permits issued today in the United States require a formal reclamation plan and some form of guarantee (i.e., financial assurance) that reclamation will be done. Forward-looking quarry operators plan well in advance for the post-quarrying use of the site. Closed pits or quarries can be reclaimed as natural habitat, especially if they intersect ground water. Other uses include golf courses and other recreation, residential or commercial development, and parks. In many cases, post-closure uses equal or exceed the value of the pre-quarry use.

Wisely restoring the environment after aggregate production requires a design plan that responds to a site's physiography, ecology, function, artistic form, and public perception. Operating and reclaimed pits/quarries are no longer isolated from their surroundings. Analysis of a pit/quarry must go beyond the site-specific and relate to its context in the regional environment so a sustainable approach is useful. Understanding this aesthetic turns an industrial site typically perceived by the public as being undesirable into a positive feature for the entire region.

Aggregate extraction site on the north end of Velarde that created an unstable highwall. Such sites can create problems for nearby communities including dust, truck traffic, infrastructure damage, noise, and safety hazards. The back side of this highwall is currently being mined, which will eliminate the steep slope and allow revegetation so the site will more closely resemble the surrounding terrain.

CHAPTER THREE

POLICY, ECONOMICS, AND THE REGULATORY FRAMEWORK

DECISION-MAKERS
FIELD CONFERENCE 2005
Taos Region

Tailings ponds west of Questa, October 2002.

An Industry Perspective on Mining in New Mexico

Patrick S. Freeman, *St. Cloud Mining Company*

The extraction of natural resources has been a historical and cultural way of life in New Mexico and, including income from the Permanent Fund, contributes as much as 35 percent of the state's annual budget. Oil and gas make the biggest contribution to the Permanent Fund, but all extractive industries, including metal mining, industrial minerals, and sand and gravel, pay severance and resource taxes, which accumulate in the Permanent Fund and finance education. Most people understand that mineral production in various forms is necessary to maintain our standard of living and provide the raw materials necessary to sustain the quality of life we all enjoy. On the other hand, participants in the reform process include groups who, on one side, believe that no extractive industry whatsoever is acceptable and, on the other, believe that there should be no regulatory constraints of any kind. The reasonable objective should be to assure that environmental concerns are met in a manner that encourages sustainable mineral development. With regulatory oversight and modern reclamation technology, mining and environmental responsibility do not have to be mutually exclusive.

The two laws that contain the most rigorous requirements on mining activities in the state are the Water Quality Act and the Mining Act. Ground water quality impacts at mining operations are regulated under the Water Quality Act, administered by the New Mexico Environment Department. The department requires mine closure plans, water treatment and contamination abatement where applicable, with financial assurance, for operations expected to impact ground water and surface water. For these operations, Environment Department requirements are usually the most challenging for mining operators to meet. This includes the largest operations in the state. For mines that do not have significant water issues, the Mining Act reclamation requirements are likely to be the most onerous. This category includes most medium-sized and small mines in the state.

THE NEW MEXICO MINING ACT OF 1993

The New Mexico Mining Act is a Governor Bruce King legacy and was carried by Gary King through an eleventh hour legislative process in 1993. Reform was long overdue, and many mining companies had failed to become proactive in their communities or were unable to address citizens' concerns. Although the New Mexico Environment Department had some authority over the most blatant abuses of air and water quality, for the most part, the hard rock mining industry had operated with virtual impunity from reclamation since Territorial times. Local groups were insisting on public oversight, and mining companies were an easy and popular target.

The poster child for mining reform was the Ortiz mine, an inactive gold operation near Cerrillos, with a scar visible from the state capitol and cyanide contamination reported in downstream ground water. Santa Fe County reacted with a mining ordinance that became a model for the state Mining Act ten years later. Elsewhere in New Mexico, environmental activists and citizens were concerned about public safety and were reacting to hazardous mine openings, abandoned mine sites, pollution, and the possible effect of some operations on general quality of life issues.

The resultant Mining Act of 1993 changed the way mining is done in New Mexico and continues to evolve administratively and through interpretation. The act was initially intended to assure reclamation of depleted or abandoned mines, but now it materially affects all phases of permitting from exploration through mine closure and beyond. Excluded from regulation under the act are sand and gravel operations, certain uranium processing, all of the coal and potash industries, copper smelters, and recreational prospecting. Some of these industry segments are regulated under other federal, state, and public land requirements. Included under the act are primarily mining and exploration activities for base and precious metals and a wide range of industrial mineral operations.

Existing mines in 1993 were allowed to continue operations if they were able to conform with the provisions of the act including achieving an approved post-mine land use and providing the state financial assurance in the form of bonds or other irrevocable security to assure future reclamation. The act holds mine operators responsible for public and private lands and was retroactive for any mines operating for at least two years since 1970, which brought many inactive and abandoned mines sites under the state's control.

The regulatory process for bringing a new mining operation into production or for expanding an existing mine is much more restrictive and uncertain. Requirements include oversight from various regulatory agencies, environmental studies, public participation, and a lengthy process of hearings, options for appeals, and possible lawsuits, some of which the applicant pays for. A few companies have tried, but in the final analysis, there has not been a new large mine operation or a significant expansion of an existing mine since the act was promulgated in 1993. Organized exploration efforts for the regulated commodities, which is a precursor to future mineral development, is virtually non-existent in New Mexico and will not likely be resumed unless there is a reasonable certainty that a potential discovery can be developed. Mining companies evaluate cost and risk and will ultimately take the path of least resistance. For now, that path leads away from New Mexico. The mines already in production, when unable to expand or respond to changing conditions, are more vulnerable to competition, increased costs, and fluctuations in commodity prices.

In the decade before the Mining Act, ten or more operations were significantly expanded and new mines were brought on line. That decade saw the reopening of the Hurley copper smelter and the Continental mine, the SXEW investments at Chino and Tyrone and at a number of new, smaller base and precious metal mines, and exploration projects scattered across the state. Pinos Altos, Carrache Canyon, mines in the Lordsburg area, St. Cloud's silver mines, and exploration projects collectively provided economic development in many rural settings. By the mid-1990s the industry experienced a downturn in metal prices, and the added costs of the Mining Act and other regulatory impacts were too much to overcome. In 1999, for example, New Mexico lost 1,426 high paying jobs and another $52 million in annual payrolls from the copper industry alone. Many of these jobs, including the production equipment that supported the employment, were exported to neighboring states or foreign countries where higher-grade deposits, fewer regulatory constraints, and lower costs were being offered. At the time the New Mexico mines were being closed, Phelps Dodge, for example, acquired and operated new copper mines and a smelter in Arizona and invested heavily in South American projects. The three principal copper smelters that provided market outlets for much of New Mexico's mineral production, El Paso, Hurley, and Playas, collectively remain closed.

The impact of these mine closings on local governments' ability to provide essential human services has been particularly severe in small, rural communities such as Animas, Lordsburg, and Questa. Grants, Silver City, and other cities have been able to partially diversify their economic base beyond mining but are still adjusting to reduced tax revenue and unemployment.

In the area of financial assurance for post-mining reclamation, New Mexico's requirements are generally more onerous than those of other western states. As an example, there was a long and more than $10-million process of hearings, appeals, studies, and administrative review in determining Phelps Dodge's financial assurance requirements for its Silver City operations. At one point, environmental groups were suggesting $1,200 million of irrevocable reclamation security and, although the current assurance is approximately one-third of that amount, it represents a substantial expense and ties up collateral that would otherwise be available for investment. The recently reopened Robinson copper mine in northern Nevada is similar in size, scope, and legacy impact to Chino at Silver City. Its joint financial assurance obligation with the BLM and the state of Nevada is $18,100,000 or about one-twentieth the requirement for Chino. Sumitomo, a former partner at Chino, recently paid Phelps Dodge to acquire their interest and relieve them of environmental obligations in New Mexico.

Once financial assurance is in place, there is disincentive to incur the additional expense of the actual reclamation as this constitutes double jeopardy for the operator. Further, the process for release of the financial assurance, once reclamation is completed, is subject to additional regulatory scrutiny and public input and is by no means certain. For smaller operators the costs for maintaining financial assurance and complying with reporting obligations, coupled with the annual permit fees, can be more than the actual cost of completing the reclamation. The catch is, reclamation can't be completed until mining ceases.

As another example of the economic impact of the Mining Act on operators, St. Cloud mines zeolite, an inert absorbent material involving no chemical processing, at a small and remote site in southwestern New Mexico and is subject to the act. Essentially all of the reclamation that can be done, short of closing the mine, has been completed. Maintaining financial assurance, annual permit fees, and administrative oversight for reporting constitutes a 4-6 percent burden over gross operating costs, not including the actual reclamation that is routinely done. St. Cloud's principal competitor in Texas has no annual fee or reclamation assurance obligations whatsoever. Although this could change, New Mexico operators are currently at a disadvantage.

Enforcement and administration of the Mining Act is under the Mining and Minerals Division (MMD) Reclamation Bureau, consisting of about ten employees. The bureau is responsive and professionally staffed. The act, however, was not funded by the legislature and derives its operating budget from the regulated operators in the form of permitting fees, annual assessments, and penalties. With a declining base of operators, as existing companies fall victim to depleting reserves and increasing costs, and as reclamation is completed and more inactive mines are released, the pool that funds the Reclamation Bureau is decreasing. The bureau's operating cost, on the other hand, has more than doubled since inception in 1993, and the remaining mines, bearing the brunt of these costs, find it increasingly expensive to remain open.

In cases such as San Pedro, Carrache Canyon, Little Rock, and others, MMD either issued permits or was unable to proceed with the permitting process because other state agencies, such as the Environment Department, or local ordinances, as in Santa Fe County, blocked or hindered the process. In other cases, as with Copper Flat near Hillsboro, citizens' groups filed appeals and demanded more and more studies until the operator succumbed to the process and withdrew from the state. In 1999 Chapman, Wood and Griswold, a consulting firm in Albuquerque, was asked to review opportunities for a possible wallboard and soil amendment operation that needed a gypsum resource and the availability of natural gas and transportation facilities, all of which are plentiful in New Mexico. The company envisioned a $60,000,000 project with employment for 60–100 workers. The consulting firm outlined several likely sites, and one in southeastern New Mexico was particularly attractive. When the New Mexico regulatory component was described, the company took its search to Utah and Nevada.

Without a doubt, many good things might not have otherwise been accomplished without the Mining Act. Commendable reclamation efforts under the act include clean-ups by companies who acquired properties with abandoned or inactive mines but were not actually responsible for the prior disturbances. Lac Minerals at the Ortiz operation near Cerrillos is a prime example, but others include Rio Tinto and other owners of uranium properties near Grants and Phelps Dodge, which completed work at Pinos Altos and Hanover near Silver City and at Tererro near Pecos. St. Cloud conducted similar reclamation at San Pedro in southern Santa Fe County and at Lordsburg, including clean-up of a cyanide heap-leach facility abandoned by a previous operator. Remaining operators are now held to a higher standard, and mines are safer, cleaner, and more responsive to citizen input than they have ever been.

A few existing mines were unable or unwilling to comply with the Mining Act or the general regulatory climate and closed. Agronics, a small humate operation near Cuba, and a garnet operation at San Pedro are examples. After spending millions, other promising exploration or development projects bogged down in the permitting process and remain inactive. The gold deposit at Carrache Canyon and copper deposits at Copper Flat near Hillsboro and at Little Rock near Tyrone are examples. Other operating mines have simply succumbed to increased compliance and operating costs, including a mica mine near Velarde, which later closed as well. The state of the hard rock mining industry in New Mexico is such that capital for development or expansion is difficult to obtain and reclamation bonding and liability insurance are beyond the reach of many operators, particularly the smaller ones.

Some extractive operators have turned to commodities not regulated under the Mining Act, such as sand and gravel, and continue to expand or add new operations. A new aggregate operation at Lordsburg for railroad ballast is an example. A call for increased regulatory control of these producers by state agencies and citizen's groups was initiated in 2000, and it will likely receive additional oversight in the future.

THE NEW MEXICO ENVIRONMENT DEPARTMENT

Throughout the evolution of the Mining Act, the New Mexico Environment Department has played a key role. The Environment Department, through its Ground Water, Surface Water, and Air Quality Bureaus, administers regulations under the Federal Environmental Protection Agency, continues to permit certain aspects of mining operations and fills in regulatory gaps in the Mining Act. The Environment Department uses its considerable technical resources, large staff, and "veto" power under the act to further assure that water and air standards are met and uses this authority to leverage operators to achieve standards that were not necessarily enforceable before the promulgation of the act. For example, some closed mine operations had expired discharge plans with limited obligations for capping tailing impoundments and ground water monitoring. As a condition for approving closeout plans and financial assurance packages under the act, some operators were held to higher standards than were required under their original permits. Within the last decade the Environment Department has also begun requiring "closure plans" with full financial assurance that covers all elements of mine closure including reclamation, water treatment, and clean-up of any water contamination. Recently the Environment Department has also become much more aggressive in enforcing penalties and collecting increased fees to fund the department's expanding activities.

The Environment Department also oversees portions of the Federal Superfund program. Clean-up of the Cleveland mill tailings near Silver City and other work near Pecos, for example, was initiated by companies within Superfund guidelines. These two examples would not have been covered under the act (pre-1973), had identifiable, responsible ownership and were ineligible or too complex to be reclaimed under Abandoned Mine Land programs.

For the mining industry, the Environment Department, however, has become a cumbersome and sometimes conflicting and duplicitous impairment for environmental compliance. The Environment Department represents a bigger variable than MMD for permitting because of the more subjective rules, time and cost to process permit applications, and selective enforcement. The complexity of the permitting, reporting, and compliance process within the Environment Department requires much more expertise than a typical small operator can manage or afford.

OTHER REGULATORY OVERSIGHT AND SUPPORT

As elsewhere, the Mine Safety and Health Administration (MSHA) regulates worker safety, training, exposure limits, and related employment issues. The state Engineer's Office controls, approves, and monitors water consumption and allocation and the safety of water impoundments. There are emerging local regulatory groups, for purposes other than conventional zoning and land use. Bernalillo, Santa Fe, and Rio Arriba Counties and many tribal governments now have separate permitting and enforcement standards for extractive activities which, in general, are more restrictive or cover other operations, such as sand and gravel producers, that are not fully addressed under state authority. The Department of Game and Fish is also represented under the act and reviews endangered species, wildlife, and the revegetation aspects of mining in general before the issuance or modification of permits. The Office of Cultural Affairs reviews mining permits for possible impacts on historical or cultural sites, cemeteries, or related sensitive areas.

For the pre-1973 abandoned mining properties not covered by the Mining Act, the citizens of New Mexico are served by another group within MMD: the Abandoned Mine Land Bureau (AML). AML derives its funding from the federal Surface Mine Control and Reclamation Act (SMCRA) of 1977. Coal mine operators pay a reclamation fee of $0.35 per ton for surface mined coal and $0.15 per ton for underground mined coal, which is dispersed back to the states for worthy reclamation and safeguarding projects, including non-coal projects. The AML staff inventories, prioritizes projects, and contracts safeguarding and reclamation. To date, major projects completed by AML include abandoned and inactive coal mine sites at Sugarite Canyon, Carthage, Yankee, and Madrid and hard rock sites at Cerrillos, Oro Grande, Cochiti, Organ, Deming, and elsewhere. Additional projects funded by SMCRA are planned and include Lake Valley and additional work at Oro Grande and Sugarite Canyon. At least three of these restored sites, Sugarite Canyon, Cerrillos, and Lake Valley are, or will be, interpretive centers, parks, and nature conservatories with economic development impact through tourism.

This reclamation work may improve the public perception of mining, provide some sustainable income producing post-mine land uses, serve as a safety valve for the protection of historical and cultural sites relating to mining, and protect citizens from the hazards often found at abandoned sites.

The office of the State Mine Inspector, whose inspection duties were primarily relegated to support and accident investigations by MSHA, now provides training to new mine employees and continuing education for experienced miners. This is an especially beneficial program for smaller companies that may not have in-house training capabilities. It is not coincidental that mine safety has dramatically improved in New Mexico.

The New Mexico Bureau of Geology and Mineral Resources, a division of New Mexico Tech, has considerable professional and technical resources. It serves as repository of historical mining records, maps, and publications; provides support, information and assistance, particularly to smaller operators, regulatory agencies, and citizens' groups; and conducts research in various forms to serve the state's needs. The bureau also helps companies interested in natural resource development in New Mexico compile previous exploration information and access the latest technology, and it provides support for analytical, metallurgical, and marketing services.

OPPORTUNITIES DELAYED

Exploration, development, and utilization of New Mexico's hard rock mineral resources have gone into hibernation since enactment of the Mining Act of 1993. Many companies have decided to work in less regulated states or to accept political risk in developing countries for the uncertainties of environmental and permitting risks in New Mexico. Known deposits await future development, and exploration potential remains to be tested. Many of the abandoned properties have been brought into compliance, and many of the environmental concerns at operating mines have been corrected. The act, with Environment Department oversight, has corrected many environmental problems at mine sites, but now it serves as an effective deterrent prohibiting expansion of existing operations and the development of new projects.

Recently copper and precious metal prices have improved, and Governor Bill Richardson's administration was able to complete a closeout plan and a financial assurance package for Phelps Dodge in Silver City. This is the first bright spot for mining in many years and has allowed a partial resumption of operations with some 150 workers being recalled.

Legislative revision of the Mining Act and Environment Department regulations, perhaps with broader provisions to encompass all extractive activities within New Mexico, will be necessary before a meaningful resumption of exploration and sustainable mineral development can resume.

Economic Impact of Mining on New Mexico

John Pfeil, *Mining and Minerals Division, Energy, Minerals and Natural Resources Department*

The wealth of the world will be found in New Mexico and Arizona —Baron von Humboldt, 1803

It is difficult to overestimate the effect that mining has had on the history and development of New Mexico. The Cerrillos mining district located southwest of Santa Fe is the oldest European mining district in the United States. Activity there predates mining in the American colonies by at least one hundred years. The Spanish were mining in the area four hundred years ago, and local pueblos have been mining turquoise in the Cerrillos Hills for at least one thousand years. Mining was important even then because it was one of the few activities capable of producing a commodity valuable enough to pay the high cost of transport from remote Nuevo Mexico to Mexico City, more than 1,500 miles south. The promise of finding gold or silver deposits was one of the driving forces behind the Spanish colonization of New Mexico. Those expectations largely came to fruition during the nineteenth century and the first half of the twentieth. Since that time the trend has been decidedly down. Experts believe that there may be many factors responsible for this declining trend including world mineral economics, the lack of undeveloped high-grade deposits, globalization, and a complex regulatory environment.

Although it is clear that mining has been a major component of New Mexico's economy in the past, it is less clear what the economic impact of the mining industry in New Mexico is today. In 2003 there were 225 registered active, producing mining operations in New Mexico. This includes five coal mines, three potash mines/mills, one molybdenum mine, two major copper mines, forty industrial mineral operations, and 175 aggregate operations. Operations on Indian lands are not included, but it is known that one of the largest coal mines in the state, some industrial mineral, and many aggregate operations are located on Indian lands. Since 1998 the number of registered operations has increased by about sixty, almost all in the industrial mineral and aggregate categories.

According to the Bureau of Business and Economic Research at the University of New Mexico, there have been no comprehensive studies done to determine the total economic impact of mining to the state. However, a variety of data has been collected that highlights the relative importance of the mining sector. The data show that during the mid-1960s employment in the mining industry was about 3.5 percent of all employment (excluding agriculture) in New Mexico. (The mining category typically includes the oil and gas industry, but for purposes of this discussion, oil and gas numbers were omitted.) This percentage has been dropping ever since, with the exception of a one-year period beginning in mid-1979. In spite of the drop in the percentage of employees in the mining industry over the last forty years, the number of employees rose from approximately 9,000 in the mid-1960s to 16,500 in 1980, but has steadily declined to about 4,000 in recent years. Mining-related employment during the first half of 2004 represents about 1.8 percent of all non-farm employees statewide. Nationally, mining is less than one-half of one percent of all non-farm employment.

Wages in the mining industry are high when compared to other sectors. According to the New Mexico Mining Association, the average employee earns more than $53,000 a year, whereas the state average is less than $30,000. The New Mexico Department of Labor statistics indicate annual average mining industry wages for 2003 were $54,392. The total economic impact to the state, as determined by a 1999 study by a private consultant and based on 1998 data, is over $4.4 billion.

IMPACT OF NEW MEXICO'S BIG PLAYERS

Copper Copper was one of the first metals ever used. Because of its properties, including high ductility, malleability, and thermal and electrical conductivity, and its resistance to corrosion it is one of the most valuable metals. Copper metal is generally produced from a multistage process that includes mining, concentrating, smelting, and refining to produce a pure copper cathode. Copper is increasingly produced from acid leaching of ores.

New Mexico copper is used chiefly in the manufacture of electrical wire. Electrical uses of copper include power transmission and generation, building

POLICY, ECONOMICS, AND THE REGULATORY FRAMEWORK

wiring, telecommunication, and electrical and electronic products. Copper is readily recycled, which contributes significantly to the copper supply. China is now the world's largest consumer of copper. The price of copper has risen dramatically from $0.69 per pound in August 2002, to $0.83 per pound in August 2003, to $1.38 per pound in October 2004.

The world's second largest copper producer, Phelps Dodge Mining Company, produces copper and base metals principally in Grant County in southwest New Mexico and in Arizona. Collectively, Phelps Dodge's six mines in Arizona and New Mexico are capable of producing more than 2 billion pounds of copper annually and account for about 60 percent of total U.S. copper production. The two active Phelps Dodge operations in New Mexico are Chino and Tyrone.

In 2004 Chino Mines Company (Chino), a division of Phelps Dodge, reactivated the Ivanhoe concentrator that had been shut down since 2001. Chino reactivated mining in the Santa Rita open pit in 2003 after mining operations were idled in 2001. Production has continued at the Chino solvent extraction/electowinning (SXEW) plant throughout these periods. The Chino Smelter has been idle since 2002. With increasing copper prices in 2004 operations resumed and production increased at Chino. Employment rose at Chino from 380 employees at the end of 2003 and is expected to stabilize at slightly more than 600 employees by the first quarter of 2005.

Jobs	6,000
Average salary	$53,000
State average all jobs	$29,000
Direct economic gain	$982,000,000
Direct personal income gain	$302,000,000
Direct in-state business income	$451,000,000
Direct out-of-state business income	$19,000,000
State & local gov. revenue	$208,000,000
Total impact on NM economy	**$4,415,000,000**

Source: New Mexico Mining Association and the National Mining Associaion

Economic impact of mining in New Mexico in 2002.

SXEW production at Phelps Dodge Tyrone, Inc. (Tyrone) has remained steady during 2004, but the mining rate and employment have increased since the end of 2003. Employment at Tyrone has increased from 260 employees at the end of 2003 to about 380 employees currently, with about fifty of that number working part-time at Tyrone. Between Chino and Tyrone, approximately 400 employees will have been recalled or hired by early 2005.

Gold, silver, and molybdenum are produced as byproducts of copper processing from Phelps Dodge

MINERAL	PRODUCTION RANK	PRODUCTION VALUE ($)	EMPLOYMENT	PAYROLL ($)	REVENUE GENERATED ($) STATE	FEDERAL
Coal	12	628,291,436	1,651	110,979,081	23,612,272	10,414,900
Copper	3	158,138,070	879	26,815,001	548,521	
Gold	-	-	-	-	3,900	
Industrial minerals	-	153,198,856	663	18,708,370	941,640	
Aggregates	-	77,848,579	1,063	17,190,991	703,926	
Molybdenum	5	15,800,000	165	7,000,000		
Potash	1	202,166,863	824	47,249,963	1,456,772	2,376,622
Silver	-	-	-	-	1,763	
Uranium		-	32	1,000,000	232	
Total		1,235,443,804	5,277	228,943,406	27,269,026	12,791,522

Summary of production rank, production value, employment, payroll, and revenue for mineral commodities in New Mexico in 2003. Rank is based on quantity produced. State revenue includes royalties/rentals from state trust land mineral leases; and severance, resources excise and energy conservation tax revenues. Federal revenue includes 50% state share of federal royalties (Onshore Collections in CY 2003). Source: Production rank from U.S. Geological Survey (http://minerals.er.usgs.gov/minerals) and DOE's Energy Information Administration (www.eia.doe.gov). Employment includes direct and contract employees. State revenue data from NM Tax and Revenue Department and the State Land Office; Federal data from Minerals Management Service.

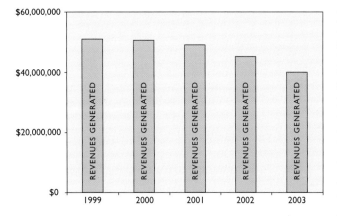

Trend information associated with several of the commodity attributes 1999–2003 including data for coal, copper, gold, industrial minerals, aggregates, molybdenum, potash, silver, sulfuric acid, and uranium. Although the figure demonstrates that the trend over the five-year period is decidedly down, with increasing commodity prices the 2004 data will likely reverse this trend.

copper operations in Grant County. Production of these commodities was highest in the late 1980s and has declined ever since. No byproduct gold, silver, or molybdenum has been produced in the state over the past several years. However, with rising metal prices and the reactivation of the Ivanhoe concentrator, byproduct production resumed in 2004.

The economic impact of the copper industry to Grant County and New Mexico was examined in a paper commissioned by Phelps Dodge in 2002. According to that document the average wage of a Phelps Dodge employee is $50,734, more than twice the average wage per job in Grant County. Wages and salaries from Phelps Dodge represent almost 8 percent of the county's total personal income. Phelps Dodge purchased $43.6 million in goods and services from other Grant County businesses and an additional $3.3 million from businesses in New Mexico but outside the county. Phelps Dodge paid $6.7 million in state and local taxes including: property taxes ($1.94 million); gross receipts/compensating taxes ($3.1 million); severance taxes on copper production ($.16 million); and fees, operating permits, licenses, and other excise taxes ($1.5 million). Total direct impact to Grant County was estimated at $80.5 million with an additional $7.8 million to other counties in New Mexico. When indirect impacts are considered, the impact to Grant County is estimated to be over $100 million and the impact to the state of New Mexico is estimated at $144 million.

Molybdenum Molybdenum is a metallic element used principally as an alloying agent in steel and cast iron. It enhances hardness, strength, toughness, corrosion resistance, and wear. These properties are achieved by combining molybdenum with other metals including manganese, nickel, and tungsten. Molybdenum is also used in the development of catalysts, lubricants, and pigments.

The state's only primary molybdenum producer is Molycorp's Questa mine and mill in Taos County, which has operated sporadically since the early 1900s. Molybdenum is also produced as a byproduct of copper production at Phelps Dodge operations in Grant County. Additional workers were hired in 2004 at Questa to meet the demands of the current market. The mine currently employs over 150 personnel with the majority of the workforce from northern New Mexico. At least 50 percent of the work force reside in the village of Questa. The average Molycorp employee earns about $50,000 annually.

According to Molycorp the mine, on an annual basis, has purchased goods, materials, and services in Taos County in the amount of $7 million and in New Mexico (particularly in the northern part of the state) in excess of $11.3 million. The direct economic benefit on Taos County alone is calculated to be more than $32 million. Taxes paid by Molycorp to the state include $141,000 in property tax; $990,000 in sales tax; and $31,000 in severance tax. Molycorp intends to continue to be a direct participant in the economic stability of the local area, including assisting the village of Questa in its economic sustainability efforts.

Potash Potash is a mined salt containing water-soluble potassium. Potassium chloride (sylvite) and potassium/magnesium sulfate (langbeinite) are mined by underground methods in Eddy and Lea Counties near

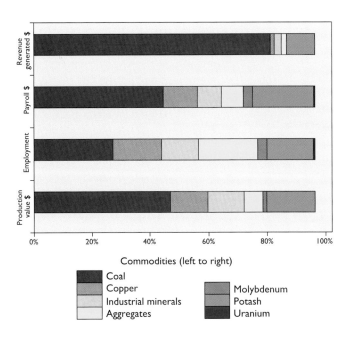

Production value, employment, payroll, and revenue generated, shown by percentage, by commodity. This figure provides some gage of the relative importance of the commodities to each other, for a selected number of attributes.

Carlsbad. More than 75 percent of U.S. potash production comes from the Carlsbad potash district, where two companies operate three mines and employ approximately one thousand workers. New Mexico potash is used primarily as an agricultural fertilizer; most is sold domestically to companies in Texas, Oklahoma, Kansas, and Nebraska. The estimated potash reserves in the district are very large. Worldwide demand is strong, inventories are low, and supply is extremely tight. Pricing has increased to historically high levels. In recent years the industry estimates their annual economic impact to the state at $304 million: $80 million on materials, energy, and related taxes; $65 million for direct payroll; $150 million for indirect payroll; and $9 million for royalties, and property and sales taxes. More than 75 percent of the total ($231 million) goes to just Eddy County, where potash employs 12–15 percent of the available workforce.

Uranium Although the present economic impact is minimal, given rising prices and huge reserves, the potential impact is huge. Since production started in the late 1940s, the Grants uranium belt has been the most prolific producer of uranium in the U.S. The boom years in the district were 1953–1980, when approximately 350 million pounds of uranium oxide (yellowcake) were produced. Uranium recovery operations declined dramatically after 1980 because of the depressed uranium market that resulted from the liquidation of large government (cold war military) stockpiles. All uranium recovery in the state ceased in December 2002. Proposed operations in New Mexico include Hydro Resources, Inc. (HRI), which intends to mine uranium by in situ leaching at Church Rock and Crownpoint in the near future. Rio Grande Resources Co. is maintaining the closed facilities at the flooded Mt. Taylor underground mine in Cibola County.

According to HRI, interest in uranium production is returning because inventories have become depleted and market prices are rising to economic levels that reflect the cost of production. The price of yellowcake has risen from about $7 per pound in 1994 to about $20 per pound in fall of 2004. The New Mexico Bureau of Geology and Mineral Resources estimates that known resources could be as much as 600 million pounds. New Mexico is second only to Wyoming in uranium reserves. HRI estimates that nearly half of the known resources are recoverable using low-cost in-situ leaching methods. In situ leaching involves the circulation of ground water with a bubbled oxygen (club soda-like) mixture through a series of injection and extraction wells that removes the uranium ore from the sandstone orebody.

According to an Albuquerque Journal article (September 1, 2003) an international consortium wants to build a billion-dollar-plus Lea County facility to produce fuel for nuclear reactors. Louisiana Energy Services (LES) announced that the $1.2 billion uranium enrichment plant would be built off NM–176, 5 miles east of Eunice, near the Texas–New Mexico border. Construction could begin within three years if the permit process goes smoothly. LES proposed moving the operation to New Mexico after encountering community resistance in Hartsville, Tennessee, where it had proposed building the facility after meeting opposition in Louisiana, its first choice. The planned Lea County facility would provide uranium for the U.S. nuclear industry, with oversight from the Nuclear Regulatory Commission and the New Mexico Environment Department. The plant is expected to employ from 200 to 400 people during construction and about 250 during operation. The company said the annual payroll will be about $10 million, with an average salary of about $50,000.

Aggregate Aggregate mining provides the basic materials for constructing and maintaining the infrastructure of New Mexico. Over 14 million tons of aggregate are consumed each year in New Mexico, with 6.5 million tons consumed in Albuquerque and Santa Fe alone. Every new home uses approximately 100 tons of aggregate, every school or hospital uses from 12,000 to 15,000 tons, and road construction requires around 8,000 tons of aggregate per mile of road. In addition to providing the basic building materials, the industry provides high paying jobs for skilled employment, funds for improving New Mexico's schools through the mining of state land, and thousands of dollars in taxes and permits. According to the National Stone, Sand & Gravel Association, every $1 million in aggregate sales creates nineteen and a half jobs, and every dollar of industry output returns $1.58 to the economy.

Construction sand and gravel is one of the most accessible natural resources and a major basic raw material. It is used mostly by the construction industry. Despite the low unit value of its basic product, the construction sand and gravel industry is a major contributor to and an indicator of the economic well-being of the state and the nation. According to the U.S. Geological Survey, New Mexico is a significant producer of construction sand and gravel and dimension stone.

Based upon preliminary U.S. Geological Survey data, New Mexico was twenty fifth in rank (twenty fourth in 2002) among the fifty states in total non-fuel mineral production value and accounted for nearly 1.5 percent of the U.S. total.

Can we expect the future to look like the past? How does economics of mining relate to sustainable development? Unfortunately, like many of life's questions, there are no clear answers. What is clear is that to date there has been no comprehensive, impartial assessments performed and, without them, economic impact is in the eye of the beholder. The highly paid mine worker or the profitable mine operator will have one view of economic impact, whereas the environmentalist, the resident living next to an aggregate operation, or the tourism-related businessman may have another.

With recent price increases there is reason to believe that the industry may make a comeback. Unfortunately, the boom and bust cycle of the mining industry offers no guarantees that the comeback will last. Our state is lucky in one regard: we possess and produce significant quantities of aggregate, particularly sand, gravel, and crushed rock. Despite the low

COMMODITY	WHERE LOCATED	USES	COMMENTS
Perlite	Rio Arriba, Socorro	Building construction products and horticultural aggregate	New Mexico is first in production. Most shipped to west coast. Greece is primary competitor.
Gypsum	Bernalillo, Sandoval	Wallboard (typical home contains more than 7 metric tons)	Centex is major player.
Pumice	North central NM	Building products	New Mexico is third in production.
Humate	Sandoval, McKinley	Soil conditioner	Significant reserves exist.
Salt	Eddy	Feedstock and highway deicing	
Zeolite	Sierra	Pet litter, animal feed, horticultural applications	New Mexico has largest zeolite mine in U.S.; first in production.
Mica	Rio Arriba, Taos	Joint compound, paint, roofing, drilling additives, rubber products	New Mexico is third in production
Limestone	Bernalillo	Cement	Tijeras plant.
Clay	Northern NM (common clay) Luna and Grant (fire clay)	Adobe brick, brick, roofing granules, and quarry tile	Fire clay is quarried for use in the copper smelter.
Iron ore	Magnetite tailings in Grant County	Used in steelmaking and to increase the strength of cement	New Mexico is third in usable iron ore production.
Sulfuric acid	Grant County	Copper recovery and a multitude of industrial processes	Produced as by-product of copper smelting.
Sand and gravel	Generally along Rio Grande corridor	Road and building construction	Represents 35% of all aggregate production.
Crushed stone	Generally in eastern and western New Mexico	Road and building construction	Represents 35% of all aggregate production.

Selected information on industrial minerals.

unit value of its basic products, New Mexico's aggregate industry is a major contributor to, and indeed an indicator of, the economic well-being of our state.

The economic and statistical information presented here was generated in both the government and industry sectors. Sources include the Mining and Minerals Division of the Energy, Minerals and Natural Resources Department, which collects and publishes statistical information related to the mining industry; the U.S. Geological Survey; several of the larger operators in the state; the Bureau of Business and Economic Research at UNM; and the New Mexico and National Mining Associations.

The Economic Anomaly of Mining—Great Wealth, High Wages, Declining Communities

Thomas Michael Power, *Department of Economics, University of Montana*

Mineral extraction activities pay among the highest wages available to blue collar workers, about twice the average. In New Mexico in 2000, mineral extraction jobs paid $50,000 per year whereas the average wage and salary job paid $28,000. Given these high wages, one would expect communities that rely heavily on mineral extraction to be unusually prosperous. That, in general, is not the case. Across the United States, mining communities, instead, are noted for high levels of unemployment, slow rates of growth of income and employment, high poverty rates, and stagnant or declining populations. In fact, our historic mining regions have become synonymous with persistent poverty, not prosperity: Appalachia (coal), the Ozarks (lead), and the Four Corners (coal) areas are the most prominent of these. Federal efforts have focused considerable resources at overcoming the poverty and unemployment found in these historic mining districts. In addition, the Iron Range in Minnesota, the copper towns of New Mexico, Michigan, Montana, and Arizona, the Silver Valley of Idaho, the gold mining towns of Lead and Deadwood, South Dakota, etc. are also not prosperous, vital communities. Over the last several decades some of these areas have begun to recover as a result of the immigration of new, relatively footloose residents and economic activities, but that recovery is entirely non-mining based.

The dramatic contrast between the wealth created and the high wages paid in mining and the poor economic performance of mining communities needs to be understood before expanded mineral extraction activities can be safely promoted as a local economic development strategy. This paper looks at the actual performance of mineral communities over the last quarter century and then turns to an explanation for that relatively poor performance.

CONTEMPORARY AMERICAN MINING COMMUNITIES

In order to explore the contemporary local impact of reliance on mining in the United States, we studied the economic performance of all U.S. counties where mining (excluding oil and gas extraction) was the source of 20 percent or more of labor earnings between 1970 and 2000. There are about one hundred such counties that could be identified out of the 3,100 counties in the U.S. Data disclosure problems prevented the identification of some mine-dependent counties.

The U.S. mining-dependent counties are spread out over half of the American states but are geographically clustered in the Appalachian (Pennsylvania, West Virginia, Tennessee, Kentucky, and Virginia) and Mountain West states. The century-old copper mines of Upper Michigan, Montana, Utah, Arizona, and New Mexico are included, as are the new gold mines in Nevada. The older coal mines in southern regions of the Great Lakes states (Illinois, Indiana, and Ohio) are included, as are the new open pit coal mines of Wyoming, Montana, Utah, Colorado, and New Mexico. The lead mines of the Ozarks in Missouri, the precious metal mines in the Black Hills of South Dakota and the Silver Valley of Idaho, and the iron fields of Minnesota are also included.

The question we sought to answer was whether this high degree of reliance on mining allowed these counties to outperform those counties that did not rely heavily on mining. For those counties that were dependent on mining in the 1970s, we looked at their economic performance in the following decades: 1980–1990, 1990–2000, as well as the two decade period, 1980–2000. For those counties that were dependent on mining in the 1980s, we looked at their economic performance in the 1990–2000 period. Economic performance was measured in terms of the growth in the aggregate labor earnings of residents of the county, per capita income, and population. In addition, the level of per capita income at the beginning and end of the periods was analyzed.

The decade of the 1980s was not a good one for mining-dependent counties. Labor earnings in those counties grew much more slowly than in other counties, almost 60 percent slower. During the 1990s earnings were still growing more slowly in mining-dependent counties, 25–30 percent slower. For the entire period 1980–2000, aggregate earnings in mining-dependent counties grew at only half the rate of other American counties.

Per capita income also grew more slowly during the 1980s in mining-dependent counties, about 30 percent slower. During the 1990s per capita income grew at about the same rate as in the rest of the nation, but for the entire period, 1980–2000, per capita income grew about 25 percent slower. The *level* of per capita income was also lower in the mining-dependent counties and, given that slower growth, the gap increased relative to the rest of the nation. In 2000 the income available to support each person in a mining-dependent county was about $9,500 per year below what was available, on average, in other counties.

Most mining operations are located in non-metropolitan areas where average incomes, in general, are lower. If the mining-dependent counties are compared only to other non-metropolitan areas, it is still true that the mining-dependent counties have lower per capita incomes and that they have lost ground relative to other non-metropolitan counties over the last three decades. This is also true for most mining regions even if the mining-dependent counties are compared only with the other non-metropolitan counties in the same state. Of the twenty five states with mining-dependent counties, only four (Montana, Minnesota, Michigan, and Georgia) had per capita incomes in the mining-dependent counties above the state's non-metropolitan average, and those incomes were only 3–11 percent higher. Of those twenty five states with mining-dependent counties, nineteen saw per capita income in the mining-dependent counties deteriorate relative to the state non-metropolitan average between 1980 and 2000.

Given this poor economic performance in U.S. mining-dependent counties, it is not surprising that population growth in these counties was negative during the 1980s and significantly slower than in the rest of the nation in the 1990s. For the 1980–2000 period, population growth in mining-dependent counties was only one-fourth to one-eighth of what was found on average in the other U.S. counties.

It is clear that over the last several decades, dependence on mining did not allow U.S. communities to perform better than other American communities. In fact, mining-dependent communities lagged signifi-

	LABOR EARNINGS			PER CAPITA INCOME		
	1980–1990	1990–2000	1980–2000	1980–90	1990–2000	1980–2000
Mining-dependent counties in 1970s	0.41	0.75	0.49	0.71	0.97	0.77
Mining-dependent counties in 1980s	0.41	0.69	0.46	0.72	0.95	0.76

Source: REIS CD-ROM; author's calculations

Growth in labor earnings and per capita income, mining-dependent relative to other U.S. counties.

	POPULATION GROWTH			LEVEL OF PER CAPITA INCOME		
	1980–90	1990–2000	1980–2000	1980	1990	2000
Non-mine-dependent counties	4.5%	11.2%	18.1%	$ 10,201	$ 19,622	$ 29,548
1970s mining-dependent	-3.0%	6.8%	4.6%	$ 8,362	$ 13,595	$ 19,893
1980s mining-dependent	-3.8%	5.5%	2.2%	$ 8,390	$ 13,754	$ 20,099
Difference: 1970 mining-dependent and other counties	-7.6%	-4.4%	-13.5%	$ (1,839)	$ (6,027)	$ (9,655)
Difference: 1980 mining-dependent and other counties	-8.3%	-5.6%	-15.8%	$ (1,813)	$ (5,874)	$ (9,457)

Source: REIS CD-ROM; author's calculations

Population growth and level of per capita income, mining-dependent and other U.S. counties.

cantly behind the average for the rest of the nation.

These are not new results. U.S. Department of Agriculture analyses have also pointed out the slower economic growth and lower per capita incomes in mining-dependent counties. In addition, a recent report by the U.S. Census Bureau providing *Profiles of Poor Counties* showed, when counties are classified by the type of industry that dominates the local area, mining counties had the highest poverty rates of any industrial group and that poverty rate increased systematically between 1989 and 1996.

Alabama	1.05
Arizona	2.64
Colorado	1.31
Illinois	1.50
Indiana	1.38
Kentucky	1.64
Montana	1.76
New Mexico	1.38
North Dakota	1.82
Ohio	1.75
Pennsylvania	1.44
Texas	1.23
Utah	1.73
Virginia	2.95
West Virginia	1.27
Wyoming	1.02
All U.S. coal counties	**1.55**
Source: U.S. Department of Labor; author's calculations	

Ratio of the unemployment rates in U.S. coal counties to the statewide average unemployment rate, 1990–2000.

Unemployment is also higher in mining-dependent counties in the U.S. For instance, unemployment rates in coal-mining counties are significantly above the average unemployment rate in the state where the county is located. Averaged over the 1990–2000 period and across all coal-mining counties, the unemployment rate in those counties was 55 percent above the state average rates. For some states, such as Arizona and Virginia, the coal county unemployment rates are two to three times higher than the state unemployment rates. Given the ongoing job losses in most coal mining counties due largely to labor-displacing technological change, these high unemployment rates might be expected. During the 1980s, for instance, the layoff rate in the mining industry was the highest of all the major industrial groups in the U.S., and the rate of job displacement in coal mining was much higher than in mining as a whole.

The important point to be drawn from all of these statistical results from an economic development perspective is that whatever might be said about the impact of mining on *national* economic development, in the U.S. these mining activities, in general, have not triggered sustained growth and development in the *local* regions were the mining took place. Closure of the mines often led to "ghost towns" and abandonment of the region. Where mining persisted over longer periods, it did not trigger a diversification of the economy. Instead, as labor-saving technologies reduced employment opportunities, the region around the mines became distressed with high unemployment and poverty rates. This was not just a historical problem associated with nineteenth-century mineral developments on the American frontier. Contemporary American counties that depend on mining continue to experience the same results, lagging the national economy.

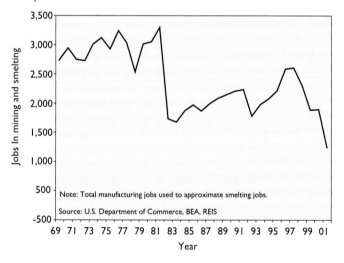

Mining and smelting employment, Grant County, New Mexico.

EXPLANATIONS FOR THE POOR ECONOMIC PERFORMANCE OF MINING COMMUNITIES

The explanation for this poor economic performance despite the local economy's specialization in a very high wage industry lies in the instability of employment and income associated with mineral development activity. As the experience of the Silver City area of Grant County, New Mexico, documents, mineral development almost always has a boom-bust aspect to it that is tied to the wide fluctuations in world commodity prices. By 2001 almost two-thirds of the mining and smelting jobs that existed in 1981 in Grant County had disappeared. When mining-related jobs represent almost a third of all local jobs, such fluctuations in employment can have a devastating impact on the community.

In addition, technological change is continually reducing the number of jobs associated with any given level of mineral development. The productivity record, for instance, in copper mining over the last quarter century is indicative of the mining industry as a whole. In copper, output per worker has tripled.

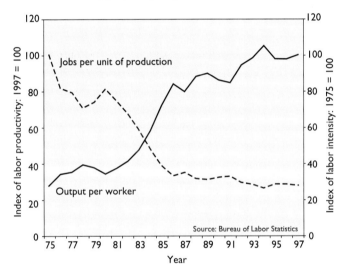

Copper mining productivity and labor intensity.

This has helped copper mining companies control costs and remain competitive while processing lower and lower grade ores. The downside of this growth in labor productivity for workers and communities is that the labor required per unit of production has been cut to a third of what it otherwise would have been. Thus, even if production is stable, employment continues to fall. Only constantly expanding mineral development can maintain stable employment, and this is never possible over the long run. Another reason for declining employment and earnings in mining is that mineral deposits are always, ultimately, exhausted, and the industry has to shift to new geographic areas. In addition, because of the high profits that are often associated with extracting gifts of nature, there tends to be ongoing struggles between miners and mining companies over the sharing of those rents. This has led to often bitter and extended strikes and lockouts that have also taken their toll on local communities, adding still another source of economic instability. Finally, mineral extraction tends to be land-intensive, imposing a disruptive footprint on the natural landscape and contributing to significant environmental degradation. This makes mining-dependent areas less attractive places to live, work, and do business, depressing economic diversification and development.

These well-known explanations for economic instability in mineral-dependent economies lead investors to be very cautious about the investments they make in areas dependent on mineral production. Since workers, residents, businesses, and local governments do not know how long the employment and payrolls will last, they reduce their risk by avoiding fixed investments that may be lost if the mineral industry enters a period of decline. As a result, mineral workers commute long distances to jobs, maintaining residences at some distance from the mineral development. Businesses are hesitant to develop local commercial infrastructure, and local governments are hesitant to finance public infrastructure with debt. The result is a less fully developed local economy and more income leakage out of the local economy. In short, excess dependence on mineral development tends to constrain local economic development, leading to the depressed economic conditions that have come to characterize many mining-dependent areas.

The policy implications of this description of the problem are straightforward. Continued dependence on one industry is probably not a good economic development strategy. Diversification away from heavy dependence on mining can reduce the vulnerability of a community to the instability associated with mining. This is not to say that mining has to be abandoned. Rather, other sectors of the economy need to grow in relative terms to provide productive balance to mining. In addition, attention to reducing and then repairing the environmental damage associated with mineral extraction is important in making the community attractive to non-mineral economic activities and supporting such diversification. All of that, of course, is very easy to say but difficult to implement. Understanding the source of the problem, however, is the crucial first step in developing a solution.

Significant Metal Deposits in New Mexico—Resources and Reserves

Virginia T. McLemore, *New Mexico Bureau of Geology and Mineral Resources*

New Mexico is recognized for several economically significant and even world-class metal deposits. However, metals production in New Mexico has continued to decline since maximum annual minerals production was achieved in 1989. This decline is a result of many complex and interrelated factors, including fluctuating commodity prices worldwide and the quality and quantity of known ore in the state. Other factors that have hampered new mines from opening in the state include water rights issues, public perceptions, the state land moratorium, and the complexity, cost, and length of time required to complete the entire regulatory process in the U.S. at local, state, and federal levels. All of these factors add to the cost of mining, not only in New Mexico but throughout the world, and ultimately they determine if a deposit can be economically mined.

Metals deposits come in all types, shapes, and sizes and are based on geologic characteristics. The origin of metal deposits in New Mexico is the subject of a separate paper in this guidebook. The shape and size of metal deposits are defined by their spatial distribution in the ground, tonnage, and grade, and are the subject of this paper.

HOW DO WE DETERMINE THE GRADE OF METAL DEPOSITS?

The term *mineral resources* refers to the assessed quantity of a commodity that is known to exist or can reasonably be inferred to exist from geologic criteria. *Reserves* are well-defined quantities of a commodity in the ground that can be economically extracted at a given time and cost. *Grade* is the term economic geologists and mining engineers use to define the average composition of the commodity in the deposit.

Resources and reserves are based on geologic characteristics, probabilities, and statistics. Geologists evaluate all available information to determine if they have adequate information to determine reserves or resources. Resources and reserves of metal deposits are nearly always determined by drilling and chemical analyses (assays) of the samples obtained from drilling. Statistics are applied to extrapolate the assays over a given distance based on the spatial distribution of drill holes, resulting in the grade of the deposit. If a given deposit covers approximately a quarter square mile, one hundred drill holes in that area is better than ten drill holes. However, the cost of drilling one hundred holes is approximately ten times the cost of a program with only ten holes. Reserves and resources are the basis for determining if a deposit can be economically mined.

Companies typically report their reserves of active operations and operations in development in annual reports to the shareholders. Reserves must constantly be re-evaluated as new information on the deposit becomes available, including subtraction of the annual production. It is important to remember that reserves are only estimates of the available ore in the ground and rarely equal the actual production, for three reasons: Recovery from the mill is less than 100 percent, statistical models fail to account for the natural heterogeneities of the ore deposit, and recovery during min-

ECONOMICS OF PRODUCTION (Increasing profitability)	KNOWLEDGE OF RESOURCE OCCURRENCE (Increasing geologic certainty)	
	Inferred	Demonstrated
Economic	Inferred resources	Demonstrated reserves
Marginally economic	Inferred marginal resources	Demonstrated marginal reserves
Sub-economic	Inferred sub-economic resources	Demonstrated sub-economic reserves

Classification of identified mineral resources. Undiscovered resources are considered as hypothetical or speculative, to reflect varying degrees of geologic certainty. The determination of whether or not a resource can be classified as a reserve depends upon the economics of production at a given time.

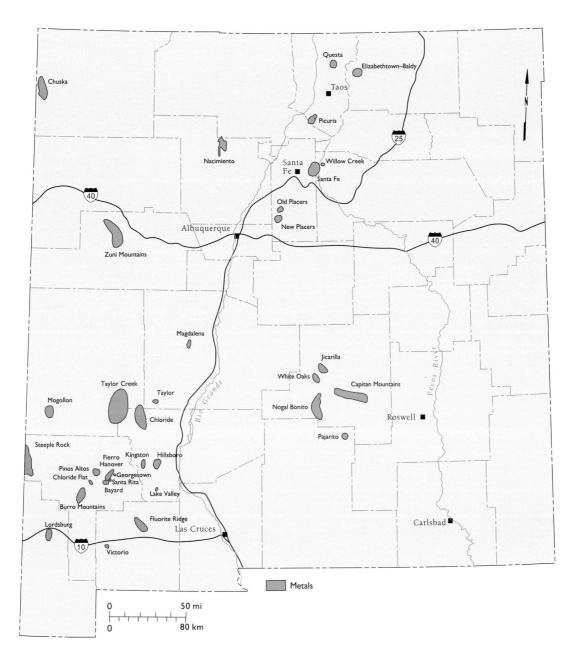

Mining districts in New Mexico with significant metal deposits.

ing is less than 100 percent.

Commodity prices fluctuate in the world economy. However, costs of exploration and mining typically increase every year. A company has to take this into account in determining if a deposit can be economically mined. Environmental and close-out costs are a major economic consideration in determining if a mine goes into production. Current predictions indicate that metal prices will continue to increase in the next decade or so as world demand for metals increases.

WORLD-CLASS AND SIGNIFICANT METAL DEPOSITS

Anomalous concentrations of copper, lead, zinc, gold, silver, and molybdenum are present to some extent in much of New Mexico and in many other places in the world. A *mineral deposit* is any occurrence of a valuable commodity or mineral of sufficient size and grade (concentration) to allow for economic development under past, present, or future favorable conditions. An *ore deposit* is a well-defined mineral deposit that has been tested and found to be of sufficient size, grade, and accessibility to be extracted (i.e., mined) and processed at a profit at a specific time. Thus, the size and grade of an ore deposit changes as the economic conditions change. Mineral deposits and especially ore deposits are not found just anywhere in the world. Instead they are relatively rare and depend upon certain natural geologic conditions to form. The requirement that ore deposits must be extracted at a profit makes them even more rare.

Statistical studies have shown that less than 5 percent of the total number of a specific type of mineral deposit produces more than 90 percent of that commodity or group of related commodities. These deposits, so-called "world-class" deposits, are the largest of the known deposits in the world in terms of size and grade. In New Mexico the copper deposits at Chino and Tyrone are considered world class. The term *significant mineral deposit* includes world-class deposits and other large deposits of economic importance today. Significant metal deposits account for most of the world's production of a given commodity. Significant metal deposits are deposits of a size and grade that would yield more than $500 million in revenues using current average metal prices. Metallic mineral deposits are typically found in large belts or provinces, but only a few of those deposits are large enough to be significant. Significant metal deposits ensure that a company can continue mining economically during bust cycles when the price is low. A company requires large deposits to meet the increasing costs of mining, refining, reclaiming, close out, and continue to make a profit for the shareholders.

Units of Measure

Industry and government in different parts of the world sometimes use different units of measure. Just to clarify:

One ton (short ton)	= 2,000 lbs
One long ton	= 2,240 lbs
One metric ton (tonne)	= 2,205 lbs (1,000 kg)

Most mining companies today use short tons (tons), but government statistics both in the U.S. and abroad are given in metric tons. The term long ton is archaic and seldom used today.

Significant metal deposits in New Mexico, as defined by the U.S. Geological Survey National Mineral Resource Assessment Team, include deposits with at least:

- 2 short tons (58,333 ounces) gold
- 85 short tons (2,479,166 ounces) silver
- 50,000 short tons (100,000,000 pounds) copper
- 35,000 short tons (70,000,000 pounds) lead
- 50,000 short tons (100,000,000 pounds) zinc

Other significant deposits found in New Mexico include:

- 1,000 short tons molybdenum
- 100,000 short tons fluorite
- 100 short tons tin
- 1,000,000 long tons iron
- 100,000 short tons manganese
- 50 short tons tungsten
- 20,000 short tons titanium

SIGNIFICANT DEPOSITS IN NEW MEXICO

Many significant mineral deposits in New Mexico will continue to attract companies to the state to develop these known deposits and explore for new ones.

Figures for reserves are only available for six deposits in New Mexico; three of these deposits are currently inactive.

At current reserves, mine life at the Phelps Dodge copper mines at Chino and Tyrone is estimated to be 5–15 years, if copper prices remain high. Resources at Copper Flat were estimated in 1995 to be 50,210,000 million tons at 0.45 percent copper and 0.015 percent molybdenum for a mine life (if this deposit goes into production) of 10 years.

Molybdenum reserves and resources at Questa (as of November 1999) were as follows:

- Proven reserves: 16,344,898 tons of 0.343 percent molybdenum sulfide at a cutoff grade 0.25 percent molybdenum sulfide
- Probable reserves: 47,198,409 tons of 0.315 percent molybdenum sulfide
- Possible reserves: 3,223,000 tons of 0.369 percent molybdenum sulfide

At current proven and probable reserves, mine life at Questa is estimated to be 20–35 years.

DISTRICT	MINE OR DEPOSIT	YEAR OF INITIAL PRODUCTION	YEAR OF LAST PRODUCTION	ESTIMATED CUMULATIVE PRODUCTION	IS THERE FUTURE POTENTIAL?	SIGNIFICANT COMMODITIES
Mogollon		1875	1969	>$25,000,000	possible	gold, silver
*Steeple Rock	Carlisle, Center, Jim Crow, Summit	1880	1993	$10,000,000	yes	gold, silver
Elizabethtown-Baldy		1866	1968	$10,000,000	no	gold, silver
*Jicarilla	Jicarilla placers	1850	1957	$165,000	possible	gold
New Placers		1839	1968	$5,750,000	unlikely	gold
White Oaks		1850	1953	$3,100,000	possible	gold, silver
*Nogal-Bonito	Rialto, Vera Cruz	1865	1942	$300,000	possible	gold, silver, molybdenum
Chloride Flat	Boston Hill, Chloride Flat	1871	1946	$13,000,000	no	gold, manganese iron
Chloride	St. Cloud	1879	1988	$20,000,000	possible	silver
Georgetown		1866	1985	$3,500,000	no	silver
Kingston		1880	1957	$6,600,000	no	silver
Lake Valley		1878	1957	$5,400,000	no	silver
Bayard		1902	1969	>$60,000,000	no	gold, silver, copper, lead, zinc
*Burro Mountains	Tyrone, Little Rock, Niagra	1879	present	>$2,000,000,000	yes	gold, silver, copper, lead, fluorite
*Fierro-Hanover	Cobre, Hanover Mountain, Continental	1889	1980	>$2,000,000,000	yes	gold, zinc, copper, iron
*Pinos Altos	Piños Altos	1860	1997	>$11,000,000	yes	gold, silver, copper, lead, zinc
*Santa Rita	Chino	1801	present	>$2,000,000,000	yes	copper, gold, silver
*Lordsburg		1870	1999	>$60,000,000	yes	gold, silver, copper, lead
Willow Creek	Pecos	1927	1944	$40,000,000	no	gold, silver, copper, lead, zinc
Nacimiento	Nacimiento	1880	1975	$1,500,000	unlikely	gold, copper
*Old Placers	Cunningham Hill, Carache Canyon, Lukas Canyon, San Lazarus	1828	1986	>$4,000,000	yes	gold, copper
*Santa Fe	Jones Hill	1956	1957	$1,000	possible	gold, silver, copper, lead, zinc
*Hillsboro	Copper Flat, Mesa del Oro	1877	1982	$8,500,000	yes	copper, molybdenum, gold, silver
Magdalena		1866	1970	$25,045,999,616	no	gold, silver, copper, lead, zinc
*†Questa	Questa, Log Cabin	1918	present	>$100,000,000	yes	molybdenum
†Picuris	Harding, Champion, Spring Gulch	1902	1955	$3,000	unlikely	tantalum, beryllium, lithium, copper
Zuni Mountains		1905	present	$5,050,000	unlikely	fluorite
Fluorite Ridge		1909	1954	$2,790,000	unlikely	fluorite
*Capitan Mountains	Capitan Mountains	1960	2000	$500,000	yes	iron
*Victorio	Gulf Minerals	1880	1959	$2,330,700	possible	beryllium, molybdenum, tungsten
*Taylor	Apache Warm Springs			0	possible	beryllium
*Pajarito	Pajarito	1952	1952	$100	unlikely	yttrium, zirconium
*Chuska	Sanostee	1952	1982	$8,000,000	unlikely	uranium, iron, titanium, zirconium, thorium
*Taylor Creek		1919	1969	$7,500	no	tin

Significant metal deposits in New Mexico, by district, based on past production and known resources or reserves. Asterisk (*) denotes those districts whose significance is primarily based on known resources or reserves. Other economic factors must be considered before mining of most of these deposits can occur. Dagger (†) denotes districts in Taos County. Production from the Chuska district has been only from uranium deposits.

MINE	MILL RESERVES (million tons)	% COPPER	% MOLYBDENUM	LEACHING RESERVES (million tons)	% COPPER
Chino	182,100,000	0.61	.02	239,000,000	0.42
Tyrone	—	—	—	252,200,000	0.31
Niagra	—	—	—	500,000,000	0.29
Cobre	57,600,000	0.55	—	77,000,000	0.26

Molybdenum and copper deposits in New Mexico. *Mill reserves* are the ore that is sent to the mill for concentration; *leaching reserves* are the ore that is produced by solvent extraction electro-winning (SXEW) of the rock piles (also known as heap leaching).

Estimates of ore in the ground or resources have been reported for additional deposits in New Mexico during the exploration phase of those deposits. These figures are not considered reserves because they are based on insufficient or uncertain information or they have not been updated to reflect economic conditions in 2004. However, these resource figures serve as an approximation of what could be in the ground.

YEAR	COPPER ($ per pound)	MOLYBDENUM ($ per pound)
1989	1.25	3.40
1990	1.19	2.85
1991	1.05	2.30
1992	1.03	2.21
1993	0.85	2.32
1994	1.07	4.51
1995	1.35	8.08
1996	1.06	3.79
1997	1.04	4.31
1998	0.75	3.41
1999	0.72	2.65
2000	0.84	2.56
2001	0.73	2.36
2002	0.72	3.77
2003	0.81	5.32
2004	1.28	

Average copper (COMEX) and molybdenum (average Platts Metals Week) prices by year, through 2004.

An Overview of the Regulatory Framework for Mining in New Mexico

Douglas Bland, *New Mexico Bureau of Geology and Mineral Resources*

Early miners did not worry about environmental issues and were not subject to governmental regulation as we know it today. Their greatest concern was simply getting authorization to explore the land for minerals, and, if they were lucky enough to find a deposit of value, to open up a mine. The Mining Law of 1872 was the first major law that established requirements that would-be miners must follow to develop a mine in the U.S., even though these requirements addressed access rather than environmental protections.

Many years passed before significant concern was raised over mining's effects on the land and its people. This eventually led to modification of the 1872 Mining Law and the passage of other laws specifically targeting certain resources such as water and air. Additional requirements were established through federal and state land management agencies, including the U.S. Forest Service, the U.S. Bureau of Land Management, and the New Mexico State Land Office, which sought to address multiple land use needs on public lands. State laws were enacted to fill the gaps or tailor protections to the needs of our state. All of these efforts are designed to protect those who may be adversely affected by mining. Such adverse impacts can be caused by land disturbance, air and water contamination, surface and ground water use, noise, dust, blasting, truck traffic, and local infrastructure use, among others.

CURRENT REGULATORY FRAMEWORK

Today the regulatory framework that applies to mineral exploration, development, and mine closure is a complicated and often confusing web of requirements established by a wide variety of entities and agencies. Most are managed by federal, state, or local government bodies and are implemented as either land management or resource protection strategies. Further complicating this issue, almost all mining operations are different, creating different issues that trigger different sets of requirements. In some cases complex sites can require years for full authorization to begin mining.

Federal Requirements

There are three main types of federal government regulations: land management, resource protection, and citizen protection. Federal lands are subject to requirements designed to allow the land to be used for mining and then returned to the government for other uses. The exception to this is certain minerals that fall under the 1872 Mining Law, which allows transfer of federal land to private citizens through the patent process. Currently there is a moratorium on patenting lands, and no lands have been so transferred in more than a decade. Both federal land management agencies, the U.S. Department of Agriculture's Forest Service and the U.S. Department of the Interior's Bureau of Land Management have regulations addressing operational and reclamation requirements that must be met as a condition of approval to initiate mining. As land management agencies, these entities have a vested interest in how the land will be used during mining, and how it will be left after mining ceases and the land is available for other uses. These requirements are broad in scope and may cover many aspects of the operation. They are likely to be the most comprehensive requirements imposed by the federal government.

Minerals on federal lands are divided into three categories: salable, leasable, and locatable. Different procedures must be followed for each type of mineral. Salable minerals include most sand, gravel, and other aggregates, and sales contracts are established that allow extraction. Leasable minerals include coal and potash, where leases are obtained and royalties paid on commodities produced. Locatable minerals, also known as "hardrock" minerals, include gold, copper, mica, and other metallic and non-metallic minerals. For these commodities, the 1872 Mining Law allows access to federal lands by staking mining claims. Regardless of mineral type, almost all mines now must have detailed mine plans that specify how and where mining will be conducted, how contamination will be avoided, and how the site will be reclaimed after mining ceases. Bonding (also called financial assurance) is required for almost all operations and consists of financial resources being provided by the operator to

FEDERAL PERMIT REQUIREMENTS FOR MINING OPERATIONS IN NEW MEXICO

AGENCY	PERMIT NAME	DESCRIPTION
U.S. Forest Service, Southwest Region 505-842-3292, and Bureau of Land Mamagement 505-438-7400	Saleable minerals: sales contract, prospecting permit, or a free use permit	Saleable minerals include common varieties of sand, gravel, stone, pumice, pumicite, cinders, and clay, however certain uncommon varieties of these minerals are classified as locatable. Sales contracts contain operational and reclamation requirements, and the payment of a percentage of sales value.
U.S. Forest Service, Southwest Region 505-842-3292, and Bureau of Land Mamagement 505-438-7400	Leasable minerals: leases, prospecting permits and licenses	Leasable minerals include potash, sodium, native asphalt, solid and semi-solid bitumen, bituminous rock, phosphate, sulfur, and coal. Leases contain operational and reclamation requirements and payment of royalties.
U.S. Forest Service, Southwest Region 505-842-3292, and Bureau of Land Mamagement 505-438-7400	Locatable minerals: notice of intent, casual use (BLM) and plan of operations	Locatable minerals generally include metallic and nonmetallic minerals not listed above as either saleable or leasable. "Significant disturbance of surface resources" triggers requirements to minimize adverse environmental impacts where feasible, and reclamation with financial assurance (bonding) is required.
Nuclear Regulatory Commission 301-415-7000	Various permits and licenses	Required for uranium milling and processing facilities including in situ leaching operations.
Army Corps of Engineers, Albuquerque District 505-342-3282	Clean Water Act Section 404 permit	Required for any discharge of dredged or fill material into waters of the United States, including wetlands, most road construction involving water crossings, and dam construction.
Environmental Protection Agency Region 6 214-655-6444	Clean Water Act National Pollutant Discharge Elimination System permit (NPDES)	Required for any discharge of pollutants from a point source into waters of the United States. "Pollutants" are defined as any material that is added to water that changes the physical, chemical, and/or biological nature of the receiving water.
Department of Labor, Mine Safety and Health Administration South Central District 214-767-8401	Requirements for health and safety of miners	A variety of requirements designed to ensure safe working conditions for miners at mine sites and on-site processing facilities. Applies to all mine types.
Department of Justice Bureau of Alcohol, Tobacco, Firearms and Explosives 505-248-6544	Requirements for authorization to use explosives	Addresses safety issues and requirements related to explosives use and blasting on the mine site.

This table contains the most common federal permits and requirements, but is not intended to be a complete list of all that may be applicable to mining operations. Not all permits and requirements listed in this table will apply to all operations. Requirements will vary depending on circumstances. Agencies should be contacted for further details. Federal requirements that are managed by State agencies through primacy programs are not in included in this table, but are listed in the following table.

the federal government to cover reclamation costs should the operator be unable to perform the reclamation. The opportunity for public input is usually included.

Resource protection laws address one particular resource, such as water quality, air quality, or endangered species. These laws apply to a wide variety of industries that may trigger their provisions. In some cases individual states have passed their own laws addressing the same resource. If an individual federal law has a primacy clause, the state may take over administration of the law if the state passes a comparable law, and the federal government often provides funding to implement the program. Primacy clauses usually require that the state law must be at least as stringent as the federal counterpart. For example, the New Mexico legislature passed the Water Quality Act, and obtained primacy for managing certain ground water quality programs covered by federal statutes.

Citizen protection laws address the health and safety of mine workers and other citizens. These include programs administered by the Mine Safety and Health Administration designed to ensure a safe work environment for miners. Another such entity is the Nuclear Regulatory Commission, which addresses issues associated with radioactive materials and radiation safety.

State Requirements

New Mexico has the same three main types of regulatory requirements as the federal government: land management, resource protection, and citizen protection. The State Land Office has land management requirements that must be followed in order to mine on state-owned lands. All minerals are managed through leases, although requirements differ depending on commodity, location, and site-specific issues. Reclamation plans are required on all leases.

The state has many resource and citizen protection laws, regulations, permits, and requirements managed by a variety of agencies. Regulated resources, materials, and facilities include surface and ground water quality and quantity, drinking water, air quality, hazardous waste, solid waste, storage tanks, cultural resources, endangered and threatened plant and animal species, and public utilities. Not all of these requirements will apply to each operation as conditions and circumstances vary widely. Some requirements are met through permits issued by an agency, such as a mining permit approved by the Mining and Minerals Division and a ground water discharge permit issued by the Environment Department. In other cases the operator must be aware of and in compliance with regulations even though permits are not issued, including cultural resource protection, worker safety, and certain endangered species requirements.

For many mining operations, the most comprehensive requirements will likely be found in the Mining Act of 1993, and if ground water is impacted, the Water Quality Act of 1967. The Mining Act focuses on what will happen to the mine property after mining ceases, and the permit outlines site-specific procedures to be followed that will establish a post-mine land use for each property. Financial assurance must be provided. The Water Quality Act requires ground water discharge permits to protect subsurface water resources. Discharge permits contain operational requirements to prevent ground water contamination during operations, and closure plans that are designed to ensure reclamation that also prevents contamination. Financial assurance may be required for this permit as well. Abatement plans may be required for existing contamination.

Local Requirements

Counties and cities also regulate certain aspects of mining operations. Many local jurisdictions have zoning laws that identify appropriate uses for specific locations, and such areas will have restrictions on certain activities. In addition, ordinances may apply that address certain activities. Controlled actions may include traffic, noise, times of operation, as well as prohibiting mining altogether in certain areas. Several New Mexico counties now have mining ordinances that specifically address if and how mining is authorized. Generally, local requirements are superseded by state and federal requirements if those requirements address the same specific areas of regulation, but local governments may "fill in the gaps" where no state or federal regulation exists.

Enforcement and Appeals

The agency responsible for implementing the regulations or establishing the permit is also responsible for enforcing them to ensure compliance. Methods for enforcement vary according to policies of the agency and the provisions outlined in the statutes and regulations. Notices of violation, fines, and corrective action plans are some of the methods used to address compliance problems.

Most decisions made by regulatory agencies can be appealed. Appeals may be made by the mining company involved or by citizens concerned with the operation. The venue for appeals varies widely, and may begin with an official within the agency itself, review by an overseeing body such as a commission or board, or it may go directly to court. Often, several additional levels of appeal are available including a State District Court, the Court of Appeals, and State Supreme Court. In some cases later appeals may be heard based on the hearing record developed at an earlier appeal; new testimony is not taken at these later appeals. If federal lands are involved, the decision may be appealed through the federal court system.

(Continues on page 110)

STATE PERMIT REQUIREMENTS FOR MINING OPERATIONS IN NEW MEXICO

AGENCY	PERMIT NAME	DESCRIPTION
ENERGY, MINERALS & NATURAL RESOURCES DEPARTMENT		
MINING AND MINERALS DIVISION		
Mining Act Reclamation Program 505-476-3400	Various mining and exploration permits	Comprehensive permits with closeout plans requiring reclamation, financial assurance and public participation. Includes minimal impact and regular permits for new and existing mines. Does not apply to aggregate and some industrial minerals.
Mine Registration, Reporting & Safeguarding 505-476-3400	Registration, reporting & safeguarding requirements	Forms required for initial registration, annual operations information, and safety methods used at closure. Applies to all commodities.
Coal Program 505-476-3400	Mining and exploration permits	Comprehensive operations and reclamation permits for all coal operations requiring post-mine land use designation, bonding, and public participation.
FORESTRY DIVISION		
Environmental Resource Assessment Bureau 505-476-3325	Scientific studies & transplantation permits	Permits for protection of threatened or endangered plant species.
ENVIRONMENT DEPARTMENT		
WATER & WASTE MANAGEMENT DIVISION		
Ground Water Quality Bureau 505-827-2918	Ground water discharge permit	Comprehensive permits requiring protection of ground water quality through monitoring, prevention of contamination, reclamation, abatement, and remediation. Required for operations that may impact ground water quality.
Hazardous Waste Bureau 505-428-2500	Hazardous waste permit and handlers permit	Required for the transportation, treatment, storage, and disposal of hazardous waste.
FIELD OPERATIONS DIVISION		
Drinking Water Bureau 505-827-7536	Construction approval for new or modified public water supply systems	Approvals address maximum contaminant levels allowed, reporting, public notification, record-keeping and water supply construction.
Radiation Control Bureau 505-476-3081	Uranium and thorium ion exchange extraction license, radioactive material license	Addresses processing and handling of radioactive and nuclear material that is generally not covered by the federal Nuclear Regulatory Commission.
ENVIRONMENTAL PROTECTION DIVISION		
Air Quality Bureau 505-827-1494	Construction and operating permits	Ensure that air pollution sources meet applicable regulations and will not exceed ambient concentration standards for air pollutants. Does not apply to Bernalillo County, which has its own program.
Solid Waste Bureau 505-827-2775	Registration, and Solid Waste Facility permit	Addresses transfer, processing, transformation, recycling, composting, or disposal of solid wastes.
Petroleum Storage Tank Bureau 505-984-1741	Tank registration, closure, investigation, reclamation	Addresses installation, operation, closure, investigation, and cleanup of sites with above-ground and underground storage tanks.
Occupational Health & Safety Bureau 505-827-4230	Miner health and safety requirements	Ensures a safe work environment in facilities not regulated by the federal Mine Safety and Health Administration. Generally applies to off-site mills and processing plants.

POLICY, ECONOMICS, AND THE REGULATORY FRAMEWORK

Agency	Permit/Program	Description
OFFICE OF STATE ENGINEER WATER RIGHTS UNIT		
Water Resource Allocation Program 505-827-6120	Permit to appropriate the public surface or underground waters of the State of New Mexico	Necessary for any person, firm, or corporation or any other entity that desires to use any of the surface or underground waters of the State of New Mexico (establish water rights). Underground permit needed only in OSE-declared water basins.
Water Resource Allocation Program 505-827-6120	Mine Dewatering permit	Ensures that mine dewatering does not impair existing water rights. Does not require or establish water rights.
Water Resource Allocation Program 505-827-6120	Approval of drill hole plugging	Approvals ensure that water encountered during drilling activities is confined to the aquifer in which it was encountered. Generally, no specific procedures have been established.
Water Resource Allocation Program 505-827-6120	License for water well drillers	Requires identification of equipment, permits for each well, record keeping, and filing of drilling log with OSE. Required for all wells drilled in declared water basins.
DEPARTMENT OF GAME AND FISH		
Conservation Services Division 505-476-8101	Scientific collection permit	Permit allows the taking of endangered wildlife for scientific and/or educational purposes.
STATE LAND OFFICE		
505-827-5750	Leases and permits for prospecting and exploration	Address operation, reclamation, and closure requirements, and includes fees and rentals. Applies to all commodities.
HISTORIC PRESERVATION OFFICE		
Office of Cultural Affairs 505-827-6320	Various permits for archaeological investigations on state-owned lands	Ensures the protection, preservation, and appropriate treatment of historic or prehistoric ruins and monuments, and any object of historical, archaeological, architectural, or scientific value. Human burial excavation permits apply to all land types.
PUBLIC REGULATION COMMISSION		
505-827-6942	Certificate of public convenience and necessity, and location permit	Required of all jurisdictional public utilities to operate, construct, or extend any plant or system, except for an extension within or to a territory already served by it.
NEW MEXICO BUREAU OF MINE INSPECTION		
505-835-5460	Underground mine diesel equipment permits, and certificates for coal miners	Various equipment and procedural requirements that must be followed to assure the health and safety of mine personnel.

Not all permits and requirements listed in this table will apply to all operations. Requirements will vary depending on circumstances. Agencies should be contacted for further details.

HOW DO I WORK WITHIN THIS MAZE OF REGULATIONS?

Miners

The first thing that must be done, no matter what your interest, is to become generally familiar with the basic requirements that apply to the activity in question and the agencies that implement them. If you are a mine operator or prospector, you will need to identify all the requirements that may apply to your proposed operation. If you are acquiring an existing operation, you should do the same thing. Permitting can be costly and time consuming. The Mining and Minerals Division offers a free publication titled *Permit Requirements for Energy and Minerals in New Mexico* that will help you get started.

Next, you should contact all agencies you believe could be involved with your project. Allow plenty of lead time, especially for those agencies that have extensive permitting or approval requirements. Discuss your particular operation with them so together you can plan a course of action to authorize mining and maintain compliance. If you suspect some of the requirements may be duplicated by another agency, ask how best to coordinate actions to minimize duplicative efforts, and ask the agencies if they can help you with this. Work hard to develop one comprehensive mining plan that meets the needs of all agencies, including those that may have overlapping requirements.

If you are aware of outside public interest in your project, meet with them early on to hear their concerns, and work with them to resolve the issues. Otherwise, you may become more familiar with regulations that deal with appeals than you might wish. Taking additional time up front to address public issues virtually always saves time in the long run, even if you don't end up in court. Above all, be patient, professional, responsive to requests for information, and question requests that you don't understand or seem out of line.

Citizens

If you are a member of the public that is concerned about an existing or proposed mining activity, many of the same recommendations listed above also should be followed. Know the regulations that apply. Develop a good working relationship with the mining company, if possible. Discuss concerns and how they might be addressed with the agency that regulates the activity in question, as agencies that do not have jurisdiction over your particular concern can do little in areas outside of their control. If it is determined the operation is in violation of certain requirements, you may want to stay involved in the remedy to ensure the problem is resolved to your satisfaction. Legal assistance, which can be costly, may be retained to help you determine the best way to work through these processes. If you wish to formally appeal certain actions, legal documents will usually need to be prepared and filed.

AGENCY	REQUIREMENT OR PERMIT
NEW MEXICO ENVIRONMENT DEPARTMENT	
GROUND WATER QUALITY BUREAU	Discharge permit-1055 for mine site
	Discharge permit-933 for tailings facility
	Discharge permit-132 for mine site sewage lagoons
	Community Right-to-Know requirements for chemical storage
AIR QUALITY BUREAU	702 Air Permit 201-M-1
RADIATION CONTROL BUREAU	Radioactive Materials License GA139-13
HAZARDOUS WASTE BUREAU	Hazardous Waste ID # NMD002899094
DRINKING WATER BUREAU	Monitoring
PETROLEUM STORAGE TANK BUREAU	Record keeping (underground fuel tanks)
NEW MEXICO STATE ENGINEER OFFICE	Tailings dams inspection & reporting
	Water consumption documenting, reporting and dam safety
NEW MEXICO MINING & MINERALS DIVISION	Mining Act existing mine permit & closeout plan
	Registration & annual reporting
U.S. ENVIRONMENTAL PROTECTION AGENCY	NPDES permit # NM0022306 for discharges to Red River
	NPDES multi-sector storm water general permit # NMR05A913
U.S. BUREAU OF ALCOHOL, TOBACCO, FIREARMS & EXPLOSIVES	Explosives and blasting requirements

The above list of permits and compliance requirements for the Molycorp molybdenum mine, Questa, New Mexico, is an example of the permits that an individual mine must obtain to operate.

Regulators

If you are a regulator, most of the above recommendations should still be followed. It is helpful to know about requirements beyond those you administer, especially those that may overlap, duplicate, or be related to yours. For these, volunteer to work with the other agency to streamline the process. It is important to educate and assist those attempting to learn about or become involved in activities you regulate. Always maintain the goal of trying to resolve the issues to everyone's satisfaction, don't just focus on a solution that meets your own agency's requirements. Make sure you include all the appropriate parties when developing solutions.

Mining has been a fundamental part of life in New Mexico for a long time, and will continue to play a significant role for a long time to come. Even if most mining in the state ceases, we will be addressing environmental and reclamation issues associated with closed mine properties for many years. Regulation of these sites has steadily increased in the past few decades because of increased concerns over the effects mines have on our world. Although a stable and unchanging suite of regulations would help miners and concerned citizens alike to understand and work within the regulatory scheme, regulatory requirements will change in the future in response to changing needs and values of society, and changing mining and reclamation methods. We should consider how proposed regulatory changes fit into the existing scheme of requirements, and what effects the proposed changes will actually have on the industry, the environment, and the public. It is incumbent on all stakeholders to consider carefully what regulations are in the best interest of all of us, not just our own interest group, now and in the future.

The New Mexico Mining Act—A Primer

Bill Brancard, *Mining and Minerals Division, Energy, Minerals and Natural Resources Department*

New Mexico was among the last states in the West to adopt a non-coal mining regulatory law when the New Mexico Mining Act was enacted in 1993. While New Mexico was late in regulating hard rock mining, the New Mexico Mining Act of 1993 is a more comprehensive and detailed law than most states. The purposes of the Mining Act of 1993 include "promoting responsible utilization and reclamation of lands affected by exploration, mining or the extraction of minerals that are vital to the welfare of New Mexico." The act establishes requirements for a broad range of "hard rock" mines to obtain permits, meet certain standards, develop an approved reclamation plan, and post financial assurance to support the reclamation plan.

WHAT IS COVERED?

The act requires all mining operations to obtain permits and meet certain requirements. Whether you are or are not subject to the act depends largely on the definitions of *mining* and *minerals* in the act. *Mining* is defined as "the process of obtaining useful minerals from the earth's crust or from previously disposed or abandoned mining wastes, including exploration, open-cut mining and surface operation, the disposal of refuse from underground and in situ mining, mineral transportation, concentrating, milling, evaporation, leaching and other processing." *Minerals* are defined as "a nonliving commodity that is extracted from the earth for use or conversion into a saleable or usable product."

The following commodities and facilities are declared exempt from the act: "clays, adobe, flagstone, potash, sand, gravel, caliche, borrow dirt, quarry rock used as aggregate for construction, coal, surface water or subsurface water, geothermal resources, oil and natural gas together with other chemicals recovered with them, commodities, byproduct materials and wastes that are regulated by the nuclear regulatory commission" and hazardous waste. Some exemptions are designed to avoid duplicative regulation of a commodity or facility already regulated under a different federal or state law, such as coal, water, oil, gas, hazardous waste and NRC facilities. However, certain commodities, such as sand, gravel, and other construction materials, caliche and potash, are largely unregulated under state or federal law. In the end, the Mining Act covers most traditional hard rock and industrial minerals including gold, silver, copper, lead, molybdenum, perlite, zeolite, silica, and garnet.

The Mining Act applies not only to all mines operating when the act was passed and to all future mines, but it also covers some mines that were no longer operating at the time the act became law. The definition of "existing mining operation" includes any "operation that produced marketable minerals for a total of at least two years between January 1, 1970, and the effective date of the New Mexico Mining Act." Therefore, a mine that produced marketable minerals for two years in the 1970s but was shut by the time the act passed in 1993 is still covered.

WHO REGULATES?

Two government entities are at the center of the New Mexico Mining Act: the Mining Commission and the Mining and Minerals Division (MMD) of the New Mexico Energy, Minerals and Natural Resources Department. The Mining Commission is charged with developing the rules necessary to implement the Mining Act and hearing appeals of permitting and enforcement actions by MMD. The Mining Commission consists of eleven members, four appointed by the Governor and seven *ex officio*. The seven *ex officio* members represent different government entities. The appointed members, consisting of two voting members and two alternates, "shall be chosen to represent and to balance environmental and mining interests."

The Mining and Minerals Division is charged with administering and enforcing the Mining Act and the Mining Commission's rules. The MMD director has broad authority under the act, including the obligation to decide on all permit applications and decide when to take enforcement action against someone violating the act or rules. The MMD director is required under the act to cooperate with other state and federal agencies that have responsibilities connected to min-

A Brief History of the New Mexico Mining Act

Gary King

In 1990 while serving as a state representative I was approached by constituents who were concerned by a new mining proposal in the Ortiz Mountains near Cerrillos. This proposal called for extraction of gold from low-grade ore via a large open-pit mine and cyanide heap leaching. Local residents were concerned about the impact this would have on their quality of life and environment. I found that New Mexico was one of only two states that did not regulate the reclamation of hard rock open pit mines. Therefore, it was virtually impossible to address the long-term impact of such mining operations through state action. This discovery led me to introduce legislation to address their concerns.

Traditionally, a bill in the legislature that impacts the environment will be referred to committees with widely differing views regarding the importance of environmental protection. Therefore, passage of the Mining Act required support from a broad range of interests and little or no stringent opposition from any major interest group.

From my perspective the primary goal of the legislation was to require hard rock miners to consider, from the inception of a project, the environmental impact of the operation on surrounding communities (primarily potential pollution of water resources with acid drainage or toxic materials) and to develop a reclamation plan to leave an economically or environmentally sound site when done. I also felt that mines that had been operating for many years might not be able to reclaim their sites to a "green field" status upon the completion, because when they were planned, they were not required to carry out their operations under such a regulatory scheme. I did not want regulations to stop mining as a viable industry within the state because of its economic importance. I also believe that reclamation of existing mine sites will generate positive economic activity in mining communities, and that industry has an obligation to protect people and the environment in areas where they are making a profit through mineral extraction.

The first piece of legislation I introduced regarding mine reclamation in 1991 was based on experience that the state had gained through the regulation of surface coal mines and from the regulation of landfills. Testimony was given at legislative hearings by experts and concerned citizens. In the final analysis, it was not possible to steer this first attempt successfully through the legislature. However, the concept was referred to an interim committee for detailed study between sessions, and a moratorium was placed on new operations within the state until the subject could be studied and a solution could be crafted.

During the summer all interested parties worked diligently trying to modify the language of the initial bill to meet our diverse needs. Hundreds of hours of work were contributed by representatives from industry and public interest groups to craft a compromise solution. We returned to the legislature, and the bill moved further through the process, but we were still unable to reach consensus with legislators. The bill failed, and everyone agreed to continue to work for another summer.

Perseverance is an important component in the recipe for success. During the 1993 legislative session virtually all of the interested parties made a commitment to find a viable solution. Our working group, which included legislators, industry representatives, and community activists, met on many evenings after the regular business of the legislature had ended. We would talk and negotiate and threaten until midnight or later. Then someone would declare that they had to get some sleep, and we would adjourn. Simultaneously, the bill was moving through the committee gauntlet, and after each hearing we would modify language to meet concerns expressed at the last hearing or to address concerns we knew would be raised by members of the next committee.

In every difficult endeavor, there is a seminal moment when success is finally achieved or calamity prevails. In our case this occurred near the end of the 1993 legislative session. Our group had reached agreement on virtually everything that was possible. Still there remained a few issues where compromise seemed impossible, primarily financial assurance. We were meeting late in the evening around the giant round conference table in the governor's office. I was surprised when Governor King came into the room and started to shake hands and slap backs of everyone in the room. He told us he appreciated all the hard work, then he left. Suddenly, he poked his head back through the door and said, "Oh yeah, one more thing, no one leaves this year until we have a bill!" That night we reached agreement on our final draft and appeared in the legislature as a unified group. Legislators appreciate having all major interest groups agree that a complex bill is acceptable, and the bill passed easily. Needless to say, the governor signed the bill too.

I have found that no complex piece of legislation can be implemented without some difficulty. One of the key provisions of the Mining Act provided flexibility to the Mining Commission to address issues such as financial assurance, considering actual conditions facing each operation. The membership of the commission is diverse, so all interested parties have a voice in the decision-making process. This flexibility has been beneficial to all concerned. The approach outlined in the Mining Act and enforced through current regulations has allowed for continuing reclamation at current sites, and we are making strides toward protection of our New Mexico communities.

ing facilities. In particular, the director must confer with the secretary of the New Mexico Environment Department on proposed rules and proposed closeout or reclamation plans for new or existing mines.

WHAT ARE THE REQUIREMENTS?

The requirements for a mining operation under the Mining Act depend on how the operation is classified. The major categories of operations are *existing mining operations*, *new mining operations*, and *exploration operations*. Within each of these categories, there is a subcategory of *minimal impact operations*. The act provides that reduced requirements should be applied to "operations that have minimal impact on the environment." Generally, minimal impact mining operations are 10 acres or less in size, and minimal impact exploration operations are 5 acres or less in size.

The Mining Act required all existing mining operations to submit a permit application by December 31, 1994, and then submit a "closeout plan" by December 31, 1995. The closeout plan is the core of the existing mine permit. The plan must demonstrate the work to be done to reclaim the permit area "to a condition that allows for the reestablishment of a self-sustaining ecosystem on the permit area following closure, appropriate for the life zone of the surrounding areas." This reclamation standard must be achieved unless the operator can show that it conflicts with an approved post-mining land use, or can demonstrate that reclamation of an open pit or waste unit "is not technically or economically feasible or is environmentally unsound." With the closeout plan, an operator must file financial assurance with the director sufficient to allow the state to hire a third party to complete the closure and reclamation requirements of the permit. An approved existing mine permit applies for the life of the operation.

The permit application process for a new mining operation is more complex. The application must contain considerable detail both on the nature and impacts of the proposed operation and on the background of the mine owners and operators. The applicant must collect at least twelve months of environmental baseline data on the permit area. The baseline investigation must provide information on (and the permit application must assure that) the operation and reclamation of the facility protect human health and safety, wildlife, cultural resources, and hydrologic balance. The Mining Commission rules require that a new mining operation employ best management practices, which include designing the operations to avoid or minimize acid drainage and other impacts to ground and surface water, to control erosion, and to use contemporaneous reclamation when practicable.

The director cannot issue a new mining permit unless he or she can find that the reclaimed operation will achieve "a self-sustaining ecosystem appropriate for the life zone of the surrounding areas" unless conflicting with a post-mining land use (no other waivers allowed), that the proposed reclamation is economically and technically feasible, and that all environmental requirements can be met without perpetual care. In addition, the operator or owners cannot fail any of the bad actor tests established under the act and the rules. A new mine permit has a maximum term of twenty years with ten-year renewal periods.

A permit for an exploration operation is the simplest to obtain. Exploration permits are valid for one year and may be renewed. Exploration operations must reclaim any disturbed areas within two years after completion of the operation.

WHAT IS THE ROLE OF THE PUBLIC?

The New Mexico Mining Act provides substantial opportunities for the public to participate in the major actions. Public notice is required on applications for the issuance, renewal, or revision of permits; for variance or standby requests; and for the release of financial assurance. The act requires that notice be provided in several manners, including mailing to all property owners within a half mile of the operation, to local governments, and to those citizens on lists maintained by MMD; posting in four conspicuous places including the facility entrance; and publishing a notice in a local newspaper. The notice provides citizens with an opportunity to comment on the proposed action and to request a public hearing.

Any person who is adversely affected by any order, penalty assessment or permit action taken by the MMD director can appeal to the Mining Commission. The commission will then conduct an evidentiary hearing on the appeal. The commission decision can be appealed to the District Court.

Finally, the Mining Act is unique among New Mexico environmental statutes in allowing a "citizen suit." A citizen with an adversely affected interest can sue any person who has allegedly violated any rule, order, or permit issued under the act, or sue the Energy, Minerals and Natural Resources Department (EMNRD), the Environment Department, or the Mining Commission for violating the act or for failing to perform any non-discretionary duty under the act.

A citizen suit cannot be commenced if the agencies have undertaken and are "diligently prosecuting" an enforcement action.

WHAT ARE SOME CURRENT MINING ACT ISSUES?

The implementation of the Mining Act over the past decade has triggered numerous disputes over the interpretation and impact of the act. Some disputes were ultimately resolved by agency actions and some by court decisions. What follows is a discussion of a few of the major issues that remain at the forefront today.

Agency Coordination

At the time the Mining Act was passed, mining operations were already subject to a number of regulatory regimes. These included water and air quality rules of the Environment Department and the federal Environmental Protection Agency (EPA), and, if the mine was on public land, the rules of land management agencies such as the Bureau of Land Management, the U.S. Forest Service, and the State Land Office. Both the environmental community and the mining community were concerned about the coordination of Mining Act requirements with those of other agencies. The environmental community feared that the various agencies would not interact and that serious concerns could fall into the cracks between regulatory programs. The mining community feared that the agencies would issue duplicative and conflicting requirements that would be costly and time consuming for the companies.

The Mining Act attempted to address the concerns of both groups. On the one hand, the Mining Act permits are treated as umbrella permits that require the operator to have obtained all other necessary permits and to obtain a determination from the Environment Department that the permitted mining activities will achieve compliance with environmental standards. On the other hand, the act also imposes requirements on all permitting agencies regulating mining operations to coordinate with each other and to avoid duplicative and conflicting requirements and permit administration.

The Mining and Minerals Division has made some strides in coordinating with other agencies. Agreements have been reached with the BLM and the U.S. Forest Service that establish processes for cooperation on mines on federal land. The State Land Office modified their rules to largely require their lessees to follow the Mining Act rules.

The greatest challenge in agency coordination has been the relationship between MMD and the Environment Department. As mentioned earlier, MMD must obtain a determination from the secretary of the Environment Department that a proposed closeout plan for a new or existing mine will achieve compliance with all applicable environmental standards. In particular, if a mining operation has potential groundwater contamination concerns, the Environment Department will require that the operator obtain a discharge permit with a "closure plan." The closure plan attempts to ensure a long-term solution to any ground water pollution concerns. Like the MMD closeout plan, the closure plan will establish reclamation requirements because in many cases the Environment Department has determined that the best way to prevent long-term water contamination is through reclamation.

While the closure and closeout plan requirements appear to establish duplicative and possibly conflicting mandates for the operators, the agencies and operators have managed to lessen the conflicts. Mine operators will often submit one plan to both agencies. The agencies will then negotiate with the operator, and amongst themselves, to establish one set of requirements for the reclamation of the facility. The agencies will also allow the operator to provide financial assurance that satisfies both permits. After the plans are approved, the agencies then continue to coordinate on the implementation and enforcement of the plans.

While the agencies have had success in avoiding duplicative and conflicting requirements, the process of agency coordination, both for state and federal agencies and for the operators, consumes considerable time and resources. The issue remains as to whether the work of the agencies can be streamlined further. Toward that end, there have been sporadic investigations into modifying the statutory framework, agency structure or staffing arrangements to combine, re-distribute or change these elements to simplify requirements and to reduce or eliminate duplication. These include combining MMD and Environment Department ground water regulatory programs, keeping them legally separate but housing them in the same location, or modifying the controlling statutes to eliminate duplicative requirements.

Financial Assurance

The Mining Act requires that each operator post, prior to obtaining a permit, financial assurance (FA) "sufficient to assure the completion of the performance requirements of the permit, including closure and reclamation, if the work had to be performed by the director or a third party contractor." The act also prohibits the operator from using "any type or variety of self-guarantee or self-insurance." These strict requirements have resulted in some of the largest financial assurance amounts in the United States: The three largest mines in the state have FA obligations that each exceed $100,000,000.

Traditionally, mines have relied on surety bonds as the primary form of FA. However, changes in the insurance industry have made surety bonds very difficult and very expensive to obtain for mining companies. As a result, both the agencies and the operators have become more creative and flexible to meet FA requirements. The Mining Commission recently amended their rules to allow additional forms of FA, including trust funds, and to allow mechanisms such as "net present value." Recent large FA submittals have included a package of instruments, including trust funds, guarantees, collateral and letters of credit. Companies have also been more willing to accelerate their reclamation work to decrease their FA obligations.

Issues still remain about the use of certain FA mechanisms. Most notable are the concerns about whether certain types of guarantees, such as those provided by parent companies, violate the Mining Act's prohibition on self guarantees.

HAS THE MINING ACT BEEN A SUCCESS?

The legislature established a goal of promoting responsible utilization and reclamation of lands impacted by mining while also recognizing that mining is vital to New Mexico. Thus, success can be measured by seeing whether the state can require responsible mining and reclamation while not killing the hard rock mining industry in New Mexico.

For existing mines, the act has, up to this point, largely been a success. Reclamation plans have been approved and financial assurance has been provided at almost all of the state's existing mines. This is a remarkable feat considering that these mines were largely developed prior to the Mining Act without any plans for reclamation. At the same time, the state's largest mines remain operational and the act has not prevented the permitting of mine expansions. The recent increases in world commodity prices has resulted in significant increases in production and employment at existing mines, and facilitated the commencement of major reclamation projects.

For new mines, the impact of the Mining Act is harder to judge. A number of new mines have received permits under the act. However, the new mines have been fairly small, and no new large metal mine has been permitted under the act. New Mexico is not alone in this regard. Few new large metal mines have opened anywhere in the continental U.S. in recent years. Metals mining is now a global industry and, for the past few decades, companies have been looking to foreign jurisdictions with large untapped high grade deposits and lower costs.

Still, some in industry will argue that the increased requirements for new mines imposed by the Mining Act and by other agencies discourage perspective mine development in New Mexico. On the other hand, environmentalists might argue that if the act prevents marginal operations from coming to New Mexico, that may also explain why New Mexico has avoided the disasters such as Summitville, Zortman/Landusky, and other mine bankruptcies, which have left most western states with considerable exposure for reclamation and environmental cleanups.

A 2004 study by the Fraser Institute, a free market think tank in Canada, offers some evidence about how the mining industry views the attractiveness of New Mexico compared to other jurisdictions. Mining executives were surveyed concerning both the policy climate and the mineral potential of various jurisdictions around the world, and their responses were used to create several indices. The institute compared the executives' attitudes about the mineral potential of a jurisdiction, both with and without their current regulatory requirements, to create a "Room for Improvement" index. The four jurisdictions with the most room for improvement were U.S. states: Montana, California, Alaska and Colorado. By comparison, New Mexico was considered to have relatively little need for improvement, finishing eleventh of the fourteen U.S. states.

Planning for Mine Closure, Reclamation, and Self-Sustaining Ecosystems under the New Mexico Mining Act

Karen Garcia and Holland Shepherd, *Mining and Minerals Division*
Energy, Minerals and Natural Resources Department

The New Mexico Mining Act (Mining Act) was passed by the New Mexico legislature in 1993. New Mexico was one of the last states in the West to pass a mine reclamation act. The New Mexico legislature wanted to address environmental issues related to mining that are not specifically covered by other statutes and create legislation that would take into account the environmental health and productivity of lands impacted by mining long after mining had ceased. The intention was to ensure reclamation of disturbed lands to a condition that provides for a beneficial post-mining use. Three of the most important concepts embodied in the Mining Act are:

- The mine area will be reclaimed so that it is environmentally stable following closure
- The reclaimed mine will support a beneficial post-mine land use
- The establishment of a self-sustaining ecosystem

These three concepts are developed as requirements in the Mining Act and drive a mine operator's planning for eventual closure and reclamation of a mining operation. Knowing these requirements exist, a mine operator can plan ahead and design the mining operation in such a way as to facilitate closure and the establishment of a post-mine land use. Future planning then becomes an essential part of the mining operation and becomes a major economic driver in a mine operator's decision to proceed with mining. *Closure* is defined as the various steps to be taken to establish a beneficial post-mine land use, while *reclamation* can be defined as the steps to be taken to stabilize the site and mitigate the impacts of mining on the environment.

RECLAMATION TO AN ENVIRONMENTALLY STABLE CONDITION

Before receiving an approved permit, a mine operator must prove to the Environment Department that he has met all state and federal environmental laws related to air and water quality. The secretary of the Environment Department will then issue a determination stating that the permit applicant has demonstrated that the activities to be permitted or authorized are expected to achieve compliance with all applicable air, water quality, and other environmental standards if carried out as described in the closeout plan. This determination will address applicable standards for air, surface water, and ground water protection enforced by the Environment Department or for which the Environment Department is otherwise responsible. The operator must also prove to the Mining and Minerals Division (MMD) that the reclaimed operation will be stable from a mass stability and an erosional point of view. Where acid-producing materials are present, the operator must address these in a way that prevents or reduces impacts to the environment. Where wildlife habitat is to be the post-mine land use, potential hazards to wildlife must be mitigated at the site.

POST-MINE LAND USE

A *post-mine land use* is defined in the Mining Act Rules as a "beneficial use or multiple uses, which will be established on a permit area after completion of a mining project" and is required by law. It is to be selected by the mine operator, who must gain concurrence from the landowner, and approval from the Mining and Minerals Division director at the time that the reclamation plan is approved. The most common post-mine land use designations for New Mexico mines are wildlife habitat, livestock grazing, and commercial/industrial. Other post-mining land use categories include cropland, pastureland, forestry, residential, recreation or tourism, water management resources, and scientific or educational. Before release of financial assurance and from further responsibilities under the Mining Act, the operator must meet criteria or standards that demonstrate that the particular type of post-mine land use proposed will be achieved at the site.

SELF-SUSTAINING ECOSYSTEM

The Mining Act requires that a mine operator establish

The Las Conchas pumice mine in the Jemez Mountains east of Santa Fe. The mining operation did not include use of chemicals or mining below ground water and was therefore more conducive to easy reclamation.

The Las Conchas mine following reclamation. The site was contoured and seeded with native vegetation so that it blended in with the surrounding ecosystem after only a few years of average rainfall.

a post-mine condition that allows for a "self-sustaining ecosystem" unless it conflicts with another type of approved post-mine land use. A *self-sustaining ecosystem* is defined in the Mining Act as "reclaimed land that is self-renewing without the need for augmented seeding, soil amendments, or other assistance or maintenance, and which is capable of supporting communities of living organisms and their environment. A self-sustaining ecosystem includes hydrologic and nutrient cycles functioning at levels of productivity sufficient to support biological diversity." Many of the post-mine land uses commonly chosen by the mine operator, such as grazing land or wildlife habitat, are compatible with the requirement to establish a self-sustaining ecosystem. This is because once the mine operator has taken the initial steps toward reclamation, such as regrading, placement of soil cover, soil preparation, seeding, and mulching, the reclaimed property should be self maintaining. After reclamation, a period of monitoring the vegetation ensues to prove to the state that the reclamation goals established in the permit have been met.

As with many hard rock mining laws across the West, the goal is to require operators to return the land to some semblance of what it was, or to create some other beneficial use, and not simply leave it as an area that has been disturbed by mining. This does not involve "restoration," which requires putting it back the way it was. It does, however, mean "reclamation," which means that the area will be stable, self-sustaining, and environmentally sound. It is understood by regulators, as well as operators, that this task is often easier said than done. Creating a self-sustaining ecosystem out of disturbed mined land can be very costly as well as very challenging.

The science of reclaiming disturbed land (reclamation science) has been in existence for over twenty-five years. It has changed and evolved over time as results of scientific studies become available. In general, reclaiming a mine site involves grading or recontouring the disturbed land, employing erosion controls on recontoured slopes, placement of topsoil or cover material, and possibly adding soil amendments such as fertilizer or organic matter. The site is then seeded and sometimes planted with woody plant seedlings. The MMD requires the use of plant species that are adapted to the site. These are typically native species, but may include drought resistant non-native species as well. A mixture of grasses, herbaceous plants, shrubs, and trees may be required to ensure a self-sustaining and diverse plant community similar to adjacent plant communities. The MMD, in consultation with the operator, establishes what the vegetation standards and goals will be for successful revegetation. Numerical standards for a percent plant cover, diversity, and density values are written into the mine permit so that the operator knows what must be achieved to meet the Mining Act requirements for revegetation. The standards are designed to meet site-specific conditions of a particular mine site. After seeding, a period of monitoring follows to determine that the site is self-sustaining. The Mining Act requires that a minimum of twelve years pass before final surveys are conducted to determine the success of the reclamation and whether the mine operator has met the permit requirements. If the post-mine land use is designated as wildlife habitat, MMD will require wildlife monitoring during the period following reclamation. The operator must demonstrate that wildlife is using the site and that there is nothing detrimental to wildlife remaining from the mining operation. To ensure the site is conducive for wildlife use, MMD encourages mine operators to leave features in place that wildlife species may use. This may include leaving a few power line poles for raptor perches and large boulder piles for small-mammal habitat.

PLANNING FOR CLOSURE

Once a post-mine land use has been chosen by the operator and approved by MMD, he can take steps that will help him achieve the post-mine land use even while the mining operation is underway. For example, if the approved post-mine land use allows for the establishment of a self-sustaining ecosystem, the mine operator must think about the best placement (location, size, and shape) and engineering design for open pits, waste dumps, leach pads, mine buildings, and infrastructure such as roads, rails, pipelines, and electrical facilities. If these mine units already exist, the operator must think about what the time frame will be for reclaiming them. Will it take one year or ten years? What units will be reclaimed first? What resources does the operator have to perform the reclamation, including man power, equipment, and financial resources? Planning ahead can save a significant amount of time and money in the long run and can increase chances of successful reclamation and eventual financial assurance release. A well-planned mining operation will involve some reclamation while the operation is still active. This is called concurrent reclamation, and it allows the operator to keep the financial assurance and overall reclamation costs to a minimum. If a mine operator waits until closure to start reclamation, it will increase the time required to obtain financial assurance release and can tie up financial resources for many years after closure.

Most mining operations require one or two years for reclamation. In New Mexico, however, some of the larger, more complex operations will take many years because of the immense size of the disturbance created by mining, often into the thousands of acres. The larger mines contain some of the most challenging environmental conditions and began operations long before the Mining Act was enacted. Often these sites require unusual or unique reclamation approaches because of steep terrain or acid-generating materials in the waste piles. In some cases acid drainage from these waste piles either is contaminating ground water or has the potential to contaminate ground water in the future. Some of these waste piles will require water treatment for many years after reclamation is complete.

While existing mines struggle to mitigate environmental impacts that occurred before the 1993 Mining Act, there are still opportunities to plan for closure, especially for new units and mine expansions. The operators should anticipate the conditions that new or expanded mine operations will create and take steps to mitigate environmental impacts before they occur.

For example, if a mine plans to locate a new waste pile or leaching facility on the mine site, it can construct the pile at the final reclamation slope angle that is conducive for plant establishment, eliminating costly double handling (regrading) of the material at a later time. This would involve designing and constructing new facilities, from the ground up, for future reclaimed slopes of 3:1 or less, instead of building slopes at angle of repose (approximately 1.3:1). Once operations at the facility are complete, the pile then would be covered with suitable growth medium and seeded. If acid drainage is expected to flow out from the pile, a liner that meets environmental standards should be placed

The Chino copper mine, a portion of which is shown here, covers over 9,000 acres and exemplifies the monumental challenges for reclamation facing some mining companies.

under the pile to ensure capture of the acidic drainage and protection of ground water. Before placement of the liner, topsoil salvage from the area can save costs associated with hauling soil cover from another source.

Many of the larger, complex mining operations are currently conducting scientific studies to determine the best way to reclaim steep slopes, acid-producing waste rock piles, acid or high-metal tailings ponds, or large open pits. The studies will be a valuable tool in determining the best approach for reclamation success. The reclamation plans for these sites are dynamic and can be changed as new information and new technology become available.

The increased awareness of modern society on environmental conservation, along with the impacts of modern mining techniques on the environment, have put new demands on the mining industry to "think outside of the pit," so to speak. Turning a mined landscape into one that will become a self-sustaining ecosystem, or other acceptable post-mine land use, takes not only a willingness to educate oneself on the science of ecology and reclamation but also a commitment to the concept of returning the land to a beneficial use once mining is complete.

Financial Assurance and Bonding: What Happens When Bankruptcy Hits

Warren McCullough, *Montana Department of Environmental Quality*

Legal mechanisms and safeguards for financial assurance should not be considered valid until actually tested and proven. As New Mexico decision makers consider the topic, it may be instructive to review the recent experience of another western mining state. Montana's plan for financial assurance before 1998, based on the conventional scientific and economic wisdom of the time and designed to shield Montana taxpayers from liabilities, received its first major test following a corporate bankruptcy, and the plan was found inadequate.

The Montana Department of Environmental Quality's (DEQ) Environmental Management Bureau (EMB) administers the state's Metal Mine Reclamation Act. Under this act, operating permits have been required for metal and stone mining operations in Montana since the state constitution was enacted in 1971. Operators are required to post performance bonds to guarantee reclamation of mine sites; reclamation standards and the language of the law have evolved over time, with the late 2004 version being:

> **82-4-338** *Performance bond.* (1) *An applicant for an exploration license or operating permit shall file with the department a bond payable to the state of Montana with surety satisfactory to the department in the sum to be determined by the department . . . . In lieu of a bond, the applicant may file with the department a cash deposit, an assignment of a certificate of deposit, an irrevocable letter of credit, or other surety acceptable to the department. The bond may not be less than the estimated cost to the state to ensure compliance with Title 75, chapters 2 [Air Quality] and 5 [Water Quality], this part, the rules, and the permit, including the potential cost of department management, operation, and maintenance of the site upon temporary or permanent operator insolvency or abandonment, until full bond liquidation can be effected.*

Performance bonds are typically submitted as surety policies, letters of credit, certificates of deposit, or cash. Please note that the act does not specifically mention corporate guarantees, and as a matter of policy, the state to date has not accepted any. From time to time until 1998, it was necessary for the state to forfeit bonds to use for reclamation of mine or exploration sites abandoned by operators, but these projects generally involved financial assurance ranging from a few hundred to a few hundred thousand dollars.

In the mid-1990s Pegasus Gold was a medium-sized gold producer, with mines in Idaho, Nevada, and Montana and a wide-ranging global exploration program. Pegasus had six mines in Montana, including four heap-leach gold operations at Beal Mountain, Basin Creek, Landusky, and Zortman, where the company had pioneered heap-leach technology in Montana starting in 1979. In 1997 the company suffered a series of financial setbacks from a steadily weakening gold price, diminished cash flow from the Montana operations due to exhaustion of permitted reserves, and, above all, disastrous losses from a new operation in Australia that failed to perform as expected. Pegasus went into Chapter 11 bankruptcy in mid-1997, booked a loss of more than $500 million for the year, and went into Chapter 7 bankruptcy in January 1998. Over the next few months, the mines that still had positive cash flow were spun off into a new subsidiary, Apollo Gold, and the state of Montana and its federal partners were handed the responsibility for reclaiming the four heap-leach properties. The insurance companies that had provided the bulk of the financial assurance for the properties in the form of surety policies had the option to carry out the work themselves, but declined. Each of the mines had unique problems.

BEAL MOUNTAIN

Beal Mountain was a 1,470-acre property with two open pits and a single 75-acre leach pad containing 15 million tons of rock at an elevation of 7,500 feet. At the time of the bankruptcy, mining operations were complete, but gold recovery was continuing. In 1997 the bond had been reduced from $11.9 million to $6.3 million in recognition of partial pit backfill and other reclamation work completed. The approved plan included treatment of the pad solution with hydrogen peroxide to break down residual cyanide, followed by land application of the treated water. After negotiations with the surety company Safeco in May 1999, DEQ received a lump sum of $6.3 million as part of a settlement agreement, which stipulated that any unused funds would eventually be returned to Safeco. The money was immediately invested in a state-controlled interest-bearing account. Then, working with a court-appointed bankruptcy trustee and the U.S. Forest Service, DEQ began to implement the approved reclamation plan. It didn't work.

When the first batch of water from the pad was treated with hydrogen peroxide and land applied, all the plants in the test area died. After extensive analysis and greenhouse testing, DEQ learned that the pad water had evolved from a simple cyanide solution into 160 million gallons of water with 1,300 ppm thiocyanate, a potent herbicide resistant to conventional cyanide treatment. Over the next several years, a whole series of unanticipated events and developments followed:

- A $1 million biotreatment system based on an analogue in British Columbia was ultimately constructed to process the pad water to reduce thiocyanate and nitrate. The system worked reasonably well, but was very temperamental and prone to "crashing."

- The sensors designed to measure water level in the pad were not calibrated properly. Days after Beal was shut down for one winter to conserve costs, the heap overflowed, creating negative headlines for DEQ and forcing expensive year-round operation.

- The high-altitude, thin-soiled land application area at Beal was not suitable for the high water application rates necessary to empty the heap before it could evolve to a more acidic condition, leading to violations of surface water standards.

- The leach pad solution did continue to evolve geochemically; the thiocyanate level decreased to a trace, while ammonia and nitrate levels increased. The biotreatment plant crashed, and DEQ actually had to buy thiocyanate to jump start the treatment process.

- Mineralized rock in place and in surface dumps was found to contribute unacceptably high levels of selenium to a local stream with a recovering westslope cutthroat trout population.

- An environmental group filed suit against the U.S. Forest Service and DEQ over the violations and the department's issuance of a discharge permit to itself.

- The suit was rendered moot when the U.S. Forest Service took the site under CERCLA (Superfund) and assumed management responsibility.

- In spite of a synthetic cap and soil cover placed over the heap, the water level in the pad is rebounding, and the treatment plant must be started once again. Miscalculation of draindown has been a common problem.

To date, long-term water treatment issues linger at Beal after the expenditure of more than double the face amount of the bond. Additional funding for the site came from interest on the bond money, gold sales shared by the trustee, millions in supplemental funding from the U.S. Forest Service (USFS), and $2.5 million in reclamation bonds sold by DEQ under authority granted by the state legislature in 2001.

ZORTMAN/LANDUSKY

Permits for mining and heap leaching of oxide gold ores on private and Bureau of Land Management (BLM) land at Zortman and Landusky in the Little Rocky Mountains were issued in 1979. As mining continued until 1990 at Zortman and 1996 at Landusky, the pits were deepened into sulfide ores, and acid rock drainage was noted around 1992. A lawsuit over water quality violations led to a consent decree among DEQ, the Environmental Protection Agency, Pegasus Gold, and its sureties before the bankruptcy, requiring the company to buy zero-coupon bonds to create a trust fund to provide for long-term water treatment after 2017.

DEQ had calculated performance bonds for earth-moving work totaling about $30 million based on the projected condition of the mines at the end of a planned and approved expansion. The price of gold fell, however, and the expansion was canceled as uneconomic. A recalculation shortly after the bankruptcy projected a shortfall of about $8 million, but the agencies received only an additional $1.05 million from the bankruptcy court while the corporate officers responsible for the company's problems received $2 million in "golden parachutes." Zero-coupon bonds that had been purchased were insufficient to create the full trust fund, but there was no company left, and the sureties took advantage of an error in the consent decree language to stop any further payments. Full funding of the trust would require an up-front investment now of more than $11 million, a sum that is simply not available to the state.

The consent decree also provided a yearly payment of $731,000 from the sureties for water treatment on site until 2017. A court-appointed site management contractor and bankruptcy trustee burned through the first year's budget in 3–4 months, leading to DEQ's dismissal of the contractor, who in turn filed a lawsuit against the agencies and a successor contractor. Six years of water treatment experience since then have shown the pitfalls of including calculations with line item amounts in agreements. Until the agreement was renegotiated in 2004, the sureties refused to pay more than the line item amount for any category in the calculation, even when other categories were underspent, and actual yearly costs ranged from $750,000 to $950,000. The total projected water treatment shortfall from the end of 2004 until 2017 is about $7.5 million.

The validity of the approved reclamation plans at the time of the bankruptcy was questioned almost immediately by tribes on the adjacent Fort Belknap Reservation and environmental groups, which ultimately led to a supplemental environmental impact statement paid for by the Environmental Protection Agency. The agencies' record of decision selected alternatives that could be largely paid for with the known funding, rather than the optimal alternatives identified. This led to another lawsuit, which lingers on, even though good engineering, additional funds from the BLM, and favorable bids from contractors have actually allowed the agencies to implement most of the optimal alternatives. In spite of reclamation to date, water quality in a drainage that flows onto the reservation continues to deteriorate. The BLM has taken Zortman/Landusky under their CERCLA authority, but there is yet another lawsuit over water quality issues to be contested.

The dirt work reclamation was largely completed by the end of 2004, but only about $2 million of the original bonds remained unspent. More than $6 million in supplemental funding has come from the BLM and state Resource Indemnity Trust grants, but significant projected long-term shortfalls remain, with no solution in sight.

LESSONS LEARNED

Lessons learned by state and federal regulators from six years of hands-on experience directing mine reclamation projects include the following:

- Site maintenance and water treatment costs continue in bankruptcy. Laws and financial assurance must be designed and written to allow regulatory agencies immediate access to funds.
- Insurance companies may prefer protracted negotiations or litigation to settlement of multi-million dollar claims.
- If reexamination of an approved reclamation plan after bankruptcy or site abandonment reveals previously unaddressed issues, the public may demand additional environmental analysis, even if there is no responsible party to pay for it.
- Financial assurance should be written to reflect involvement of federal partners.
- Financial assurance should be written to exclude line-item limitations on costs, and agencies should attempt to collect bond amounts as lump sums to be placed in interest-bearing accounts.
- It is extremely difficult in the current economic climate for even financially stable companies to obtain surety bonds. Agencies should be flexible, creative, and reasonably patient as companies try to establish acceptable guarantees for reclamation.
- Indirect costs (administrative overhead, engineering design, inflation, contingencies, etc.) are a much larger part of total reclamation costs than DEQ previously assumed.
- Real-world emergencies will continue to occur under agency management.
- The geochemistry of solutions in leach pads, tailings impoundments, and waste dumps may continue to evolve during reclamation, complicating treatment and increasing costs.
- When bond calculations include a component for long-term water treatment, DEQ runs the calculation out to one hundred years. Projected expenditures beyond one hundred years have little effect on a present-value figure.
- Bankruptcy trustees serve different masters and may sell equipment or facilities needed at the site for reclamation.
- Agencies must be creative when faced with financial assurance shortfalls. Grants or supplemental funding may be available from federal partners

(EPA, BLM, USFS). In 2002 Montana sold $2.5 million in state general obligation bonds to fund reclamation at Beal Mountain.

CORPORATE GUARANTEES

Two recent corporate histories involving prominent companies in Montana will help illustrate why the state does not wish to hold corporate guarantees for mine reclamation. For years, Montana Power Company (MPC) was a solid, secure, dividend-paying utility company. A few years ago the company divested itself of its traditional assets, including coal-fired and hydroelectric power units, transmission systems for electricity and natural gas, and oil and gas production. The proceeds of the divestitures were all plowed into telecommunications, particularly fiber optic transmission lines. That overbuilt market collapsed. The company went into bankruptcy, and the remaining assets were liquidated for pennies on the dollar. The company that purchased the transmission systems also went into Chapter 11 bankruptcy, although it has recently reorganized and emerged. A corporate guarantee from MPC for anything would have been worthless.

Stillwater Mining operates two platinum group metal mines on the JM reef, a world-class mineral deposit in Montana's Stillwater Complex. Although the stock traded in the upper $40 range only a few years ago, a free-fall drop in palladium prices and huge capital costs drove the stock down below $2.50 in early 2003. The company was widely believed to be on the verge of bankruptcy, which was averted only by a takeover and infusion of capital by Nor'ilsk Nickel, a major Russian mining company.

Such huge and sudden variations in overall value, especially in corporations perceived as solid, with substantial assets, have convinced Montana regulators of the need to avoid corporate guarantees. Had corporate guarantees been in place from Pegasus Gold, the state of Montana, with a limited industrial base and fewer than a million people, would have faced a total reclamation shortfall on the Pegasus properties alone of more than $75 million.

Financial Assurance for Hard Rock Mining in New Mexico

Ned Hall, *Phelps Dodge Corporation*

Hard rock mine operators face significant challenges in establishing financial assurance for New Mexico mining operations under the New Mexico Mining Act and the Water Quality Control Act. These challenges include obtaining approvals of the scope of work for future mine closure and reclamation from the two state agencies that administer these laws, developing and obtaining approval of cost estimates, determining the amount of financial assurance based upon the cost estimates, and establishing financial assurance mechanisms. The Chino and Tyrone mines, operated by Chino Mines Company and Phelps Dodge Tyrone, Inc., respectively, are the two largest hard rock mining operations in New Mexico, and their experiences are representative of these challenges.

CALCULATION OF FINANCIAL ASSURANCE AMOUNT

The Chino and Tyrone mines are "existing mining operations" as defined by the New Mexico Mining Act. These mines were developed, operated, and most of their current footprints in place decades before the Mining Act was enacted. Open pit mining began at Chino in 1910 and at Tyrone in the late 1960s. This long history results in special challenges, including environmental impacts that occurred before environmental regulations were established, and the application of new closure and reclamation requirements and performance objectives to facilities designed and constructed before these requirements were established. For example, these mines were designed with steep pit and stockpile slopes to minimize the footprint of the mine. Installation of soil covers as part of reclamation to reduce infiltration of precipitation and to establish vegetation requires that slopes be flattened by regrading at substantial cost.

Another major challenge is the Mining Act requirement to establish financial assurance based upon the "worst case" scenario. These worst case assumptions are that the mine operator will go bankrupt at the point in time when closure and reclamation costs are the highest, and that the state will have to hire a third-party contractor to conduct the work. Although providing maximum protection to the state, these assumptions result in financial assurance requirements that can substantially exceed the estimated cost for the operator to conduct closure and reclamation at the end of mine life.

Some mines are required to provide financial assurance for long-term water treatment. In the case of the Chino and Tyrone mines, as well as the Continental mine (final permit revision pending), the state has required financial assurance for water treatment for a period of a hundred years. It can be challenging to estimate the volume of water that will require treatment, the quality of that water, and treatment costs over such a long period. Long-term closure and reclamation plans take several years to implement and must be adjusted for cost inflation. Furthermore, the amount of financial assurance required is based upon "net present value." Initial financial assurance amounts may be reduced based upon an expectation of future growth of the principal amount over time through investment. This concept was specifically approved by the Mining Commission in changes to the Mining Act Rules made in late 2003.

For the largest mines in the state, the process of developing closure and reclamation plans, estimating the cost of conducting those plans, and determining the required amount of financial assurance took about ten years following the passage of the Mining Act. The total amount of financial assurance required of the two largest mines (Chino and Tyrone) combined exceeded $450 million. The next task was to establish financial assurance mechanisms for such large amounts.

COST AND AVAILABILITY OF FINANCIAL ASSURANCE MECHANISMS

The New Mexico Mining Act allows the use of a variety of financial assurance mechanisms, including surety bonds, letters of credit, cash certificates of deposit, trust funds, collateral, and third-party guarantees. Until the last few years, surety bonds were the mechanism of choice for many financial assurance requirements, particularly for larger operators. Surety bonds could be obtained in large face amounts by financially healthy companies for relatively modest annual premiums. Beginning in 2001 the market for surety bonds for mine closure and reclamation changed dramatically, with many insurers withdrawing from the market entirely. Consequently, surety bonds became (and remain) difficult to obtain and, when available, are much more expensive. Insurers may also require surety bonds to be secured by pledges of specific assets.

Letters of credit issued by banks can be used as financial assurance. However, letters of credit usually are issued for terms of one year or less and command significant premiums, resulting in high carrying costs. They generally are not suited to large, long-term financial assurance obligations.

The New Mexico Mining Act allows "third-party" guarantees as financial assurance. Third-party guarantees may be accepted if they are issued by corporations that meet strict financial tests designed to ensure that sufficient assets will be available to cover closure and reclamation costs. However, following the Mining Commission's 2003 amendments to the Mining Act Rules, third party guarantees now may cover a maximum of 75 percent of the total financial assurance amount for a mine.

The loss of surety bonds as a viable financial assurance mechanism, coupled with the limits on the use of letters of credit and the limitation of third party guarantees, has resulted in the need for cash and other assets to be pledged for substantial portions of New Mexico financial assurance obligations. The Mining Act Rules, passed in 1994 as a requirement of the 1993 Mining Act, did not contemplate large amounts of financial assurance being covered by cash and limited cash mechanisms to certificates of deposit subject to the $100,000 FDIC-insured limit. To provide a more suitable mechanism for larger cash deposits, the Mining Commission amended the rules in 2003 to allow for trust funds. Following this amendment, Chino Mines Company established a trust fund for one-third of its financial assurance obligation, or about $64 million, and Tyrone is obligated to provide $27 million in cash funding to a trust fund.

The Mining Act also allows for collateral, including real property, as financial assurance, as long as the real property is not within the permit area of the mining operation. Mine operators who have lands outside the permit area may prefer to pledge those assets to cover a part of their financial assurance obligation rather than cash, because the use of cash to cover financial assurance obligations precludes the use of the pledged cash for other investments, including the expansion of mining operations. The pledge of real property for financial assurance, however, has proven to be time-consuming and costly. Transaction costs have included appraisals, appraisal reviews, environmental assessments, surveys, title insurance, and other costs typical of a large real-estate transaction. Tyrone's proposal to pledge collateral for a portion of its financial assurance obligation is still in process.

CONCLUSIONS

Establishing financial assurance for New Mexico's two largest mines has been a technical, procedural, and financial challenge. As long as this process has taken, it is not yet over. The permits for these mines require additional studies of the mines, including studies of the performance of various closure and reclamation techniques and the feasibility of alternative closure and reclamation measures. The plans must be re-evaluated and adjustments may be required after the studies are completed. In addition, closure and reclamation work is now underway on inactive portions of the mines. This will require adjustments in the approved cost estimates and financial assurance required, in order to reflect the work that has been performed.

Financial Assurance—The Requirements

Douglas Bland
New Mexico Bureau of Geology and Mineral Resources

Financial assurance, also known as bonding, can be required for closure and reclamation of New Mexico non-coal mining operations by several state and federal agencies, depending on jurisdiction and potential environmental impact. State financial assurance requirements are established by the State Land Office for operations on state-owned land, by the Mining and Minerals Division (MMD) for obligations under the New Mexico Mining Act, and by the New Mexico Environment Department for addressing existing or potential ground water impacts. Federal financial assurance requirements are imposed by land management agencies if the mining operation is on land managed by either the U.S. Forest Service or the Bureau of Land Management.

The Mining Act requires that financial assurance be posted before any permit is approved for an exploration or mining operation. An exception is made for general permits and minimal impact exploration permits, mining operations that in general excavate less than 50 cubic yards of material per year, and exploration operations that disturb less than 5 acres of land. The financial assurance amount is based on third-party costs to perform the close-out or reclamation plan if the operator is unable or unwilling to perform these tasks. If the operator performs reclamation covered by the financial assurance, he may reduce the amount of financial assurance posted with MMD. Forms of financial assurance accepted by MMD include cash, surety bonds, letters of credit, collateral, trust funds, and third-party corporate guarantees.

The Environment Department may require financial assurance associated with ground water discharge permits issued under the Water Quality Act. Discharge permits are required for any operation that may have an impact on protected ground water resources, generally defined as those that contain less than 10,000 parts per million of total dissolved solids. The Environment Department generally accepts financial assurance instrument types similar to MMD. Most mining operations that are subject to financial assurance for a discharge permit are also required to submit financial assurance to MMD, even though they may be required for different aspects of reclamation because Environment Department requirements focus on water quality protection and Mining Act provisions ensure the re-establishment of a beneficial post-mine land use. These two agencies have agreements that allow one financial assurance package to be posted that meets the needs of both agencies, and they coordinate establishment of the amount, management of the financial assurance, and expenditure, if needed.

The State Land Office requires financial assurance for all types of mining conducted on state-owned lands. However, the state land office does not require duplicate financial assurance for mining activities covered by financial assurance posted with MMD.

The federal land management agencies, the U.S. Forest Service (USFS) and the Bureau of Land Management (BLM), each have three categories of mining operations. They are salable, leasable, and locatable minerals. Federal financial assurance is required for salable and leasable minerals. Both the USFS and BLM have their own regulations that apply to locatable minerals, which are further subdivided into operations that require casual use, notice level, or plan of operations permits, depending on the amount of disturbance and environmental impact anticipated. All of these require financial assurance except casual use. Financial assurance instrument types allowed are similar to those accepted by MMD, except that corporate guarantees are not acceptable. A joint powers agreement between the USFS, BLM, and MMD has been adopted to address financial assurance requirements where there is overlap between them, but conflicting requirements have prevented coordination of joint instruments between the Environment Department and the federal agencies at this time. Fortunately, there are few instances where this has been an issue. To date, no mines have defaulted under the Mining Act or Water Quality Act requiring reclamation managed by the state using forfeited financial assurance.

FINANCIAL ASSURANCE TYPE	NUMBER OF MINES	TOTAL POSTED	JOINT AGENCY AGREEMENTS
Certificate of deposit	18	$448.701	6
Surety bond	23	$88,827,858	7
Cash account	5	$82,043,304	3
Letter of credit	11	$8,693,351	4
Collateral	2	$8,564,315	2
Corporate guarantee	5	$475,011,015	4

Financial assurance posted under the New Mexico Mining Act. Joint agency agreements may be between the New Mexico Environment Department, the U.S. Forest Service, the Bureau of Land Management, or a combination of these. Data from the New Mexico Energy, Mineral and Natural Resources Department.

Will There Be Water to Support Mining's Future in New Mexico?

John W. Shomaker, *John Shomaker & Associates, Inc.*

This paper explores some factors that influence the availability of water for future mining. Allocation of water is largely a legal matter, but my perspective is that of a geologist and an observer, not an attorney; nothing that follows should be thought of as legally authoritative, nor as legal advice.

Mining enterprises always require water. The uses vary widely: dust suppression, milling and processing, conveyance of tailings from mills, recovery of metals by leaching, dewatering of underground workings, and reclamation of mined lands. Apart from simple dewatering, most of these uses can lead to relatively high depletion—most of the water is lost to evaporation, rather than returned to the surface water or ground water system.

New Mexico's mining industry (which includes oil and gas extraction) has accounted for 1.5 to 2 percent of the state's total water use, in terms of depletions (water actually lost to evaporation) in recent decades. (Published state engineer statistics do not separate out oil and gas activities.) The value of water used in mining, in terms of share of the "gross state product," is relatively high. In 2000, for example, the gross state product in New Mexico from all economic activity amounted to an average of $20,000 per acre-foot of water depleted in all uses. The mining category contributed $103,000 per acre-foot; the corresponding figure for irrigated farms was about $347 per acre-foot of water depleted.

At first glance New Mexico's water law, evidently designed to regulate water in irrigation agriculture, seems to fit the mining industry poorly. Water rights are nominally perpetual, tied to specific lands and points of diversion (either surface water diversions or wells), as long as beneficial use is made of the water. But mining is almost by definition temporary in any particular place. The water use would presumably end when the ore, coal, or industrial-mineral deposit is exhausted and reclamation activities completed. On the other hand, long dormant periods governed by changing commodity prices are typical of mining. New mining activity is commonly initiated in established districts.

Water produced for uranium mining and milling increased from zero in 1950 to a peak that may have been near 20,000 acre-feet per year in about 1980, but was about 2,600 acre-feet in 2000 and has been negligible (except for reclamation activities) since 2002. The pumping was largely for milling and for dewatering underground workings, with the water discharged to the surface drainage. The uranium industry probably over-appropriated the Bluewater Basin for some period, in the sense that depletions were sufficient to cause significant lowering of ground water levels. All of this would be primarily of historical interest, except that other users are now considering the water rights established during the uranium era, whatever they may in fact be, and uranium production itself may emerge again as energy demands continue to rise.

Copper production in southwestern New Mexico has needed water for much longer, since about 1804. Water produced for the minerals industry in Grant and Hidalgo Counties was about 8,700 acre-feet in 1962, of which about 5,300 acre-feet were depleted. The corresponding figures for 2000 were 25,800 acre-feet and 21,300 acre-feet, but between 1962 and 2000 there were major changes in the patterns of pumping from surface water and ground water sources. In Taos County, molybdenum mining and milling has required diversions of as much as 9,400 acre-feet per year (in 1976), but water requirements have varied widely depending on the status of operations. In 2002 diversions were about 2,700 acre-feet, and they have not been above 6,000 acre-feet since 1991. These kinds of variations lead to complicated questions about the real meanings of the water rights involved.

The "use it or lose it" aspect of our water law would appear to mean that a mine operator, having once acquired a water right, must find a home for it in some other beneficial use when mining ceases, or risk losing the value it represents. In actuality, many other water uses share the impermanent character of mining. Water rights for agriculture are "not necessarily" forfeited during periods of non-use when irrigated farmlands are under the acreage reserve program or conservation program provided by the Soil Bank Act, and, in general, "forfeiture shall not necessarily occur if circumstances beyond the control of the owner have caused non-use." This provision would appear to apply to mining, but there are few guidelines for its application.

Our water law does not distinguish among beneficial uses, regardless of their relative values to the community, and so would not particularly favor mining over some lesser-value use. On the other hand, high-value uses can justify the purchase of rights and the transaction costs relating to new appropriations or transfers of existing rights, whereas lower-value uses may not be able to. This may be an advantage for mining, but would not entirely offset the loss of value attributable to a long and uncertain administrative process.

The conventional permitting process of the state engineer can consume a great deal of time and seems to differ from environmental and land-use permits in that the outcome is less predictable. Compliance with regulations, however complex and burdensome, will lead to a permit from the Environment Department, but a state engineer permit may simply be denied if the applicant has failed to prove that no existing right would be impaired at some time in the future, or that the proposed project is not detrimental to the public welfare. It may not be as easy to demonstrate the future effects of ground water pumping as it is to assume an obligation to conduct an operation in some particular way. A trend toward a somewhat different policy seems to be emerging, in which the state engineer is willing to issue a permit, and somewhat more quickly than in the past, but sets conditions that represent a continuing obligation of the permittee to keep existing users whole.

Mining projects have commonly been controversial, and in some cases the state engineer administrative process, which must take the public welfare into account, has been the forum for presenting a case against a project even though the effects of its proposed water use have been small. The scope of public welfare issues is not defined, however, and may be very broad if opponents of a project are creative.

New Mexico has been engaged in regional water planning since 1987, and a statewide water plan is in the making. Unfortunately for mining, of course, as-yet undiscovered mineral deposits can't be represented in the plans, nor can the water plans anticipate price changes that would trigger reopening of former operations. One of the principal functions of the regional plans is to provide guidance to the state engineer as to each region's understanding of the public welfare. If potential mining is not specifically dealt with in a regional plan, is a new mining project at some disadvantage, simply because it could not be defined in advance and has not been examined through the public-welfare lens?

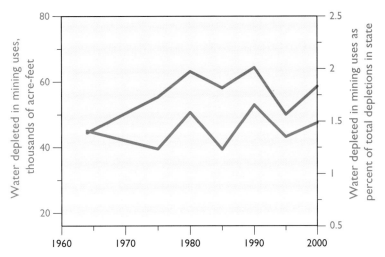

Water depleted in mining uses (including oil and gas) in New Mexico.

The state legislature in 1980 recognized that mining (uranium mining in particular) would better contribute to the economy if water could be pumped from underground workings, even though the water might not be put to any beneficial use (as required under a conventional water right), and instituted the Mine Dewatering Act. Mine dewatering was accepted as a beneficial non-use. The act actually went much further and provides for a sort of condemnation process in which any new user, not limited to mining, may take the water needed, as long as existing users' rights are kept whole through a "plan of replacement" approved by the state engineer. The Mine Dewatering Act is still in place but has been little used.

Mining commonly is distant from major streams, so that simply transferring irrigation rights upstream or downstream to serve them is less a consideration than the potential for "local impairment" in the form of increased drawdown in wells. In such a case, a plan of replacement under the Mine Dewatering Act might include providing water, or payment of the incremental increase in costs of pumping. Even though the mines themselves may be away from major streams, depletions of water still influence the amounts of water available for delivery to Texas or Arizona under interstate stream compacts, and transfer of existing surface water rights is likely required to offset these effects.

Water rights acquired for mining, either by application for a permit to appropriate water and then perfecting the right by using water, or by simply declaring the existence of a right that had been established before the state engineer asserted jurisdiction over the particular ground water basin, have an important value on a mining company's balance sheet. However, such water rights may not be as fully defined as would be desired today. Are rights perfected under permits to appropriate from the state engineer of equal status with rights that have been adjudicated in court? Are rights automatically valid for post-mining closure and environmental reclamation needs, even if those uses were not specifically described in the original permit application, license, or declaration? Do such rights have the same status and value as irrigation rights (for example) when the mining-related uses are finally at an end? Policies or litigation may be necessary to answer these questions.

If beneficial use is "the basis, the measure and the limit" of a water right (in the words of the New Mexico Constitution), what proportion of a water right would remain valid for future transfer to another use if the water requirement in the mining use declines over time? What if a large part of a right is unused for many years more than the four-year statutory period, after which the state engineer may serve a notice warning of possible forfeiture? If a water right must continue to be available for post-mining environmental uses for a long time, but water is not actually put to beneficial use, would some part of the right be deemed abandoned? How long will a water right, established or acquired for mining, continue to be valid if the mining company chooses to hold it in anticipation of reopening of the mine? And how long will the right continue to be valid if not transferred to another use?

In the San Juan Basin, many applications to appropriate ground water were filed by uranium and coal companies decades ago; some were approved and permits issued, and some are still pending, but in many cases no water has been used for years. A number of rights, established by drilling of exploration wells, or by actual operation for some period, were declared by uranium and coal companies before the state engineer asserted jurisdiction but have not been exercised for many years. The current status of these permits and rights may be described differently by people in the mining industry and on the state engineer's staff. Probably in some cases the applicant corporations no longer exist. It seems likely that the energy resources of the basin will be of economic interest again, however, and resolution of these cases will be necessary.

Much of the new mining in New Mexico is likely to be in the form of sand-and-gravel or crushed-rock operations. These tend to be near urban areas, because markets are primarily in building and highway construction, and transportation is relatively expensive. Permits relating to zoning and land use are probably more of a concern than the availability of water. Water requirements are relatively short term, governed by a typical operation's reserve life of perhaps twenty years, and with little land-surface reclamation (and therefore irrigation) required, at least for now on private lands. Requirements for reclamation, and therefore for water, seem likely to increase in the future, even on non-public lands.

CHAPTER FOUR

SUSTAINABLE DEVELOPMENT, TECHNOLOGY, AND A LOOK TO THE FUTURE

DECISION-MAKERS
FIELD CONFERENCE 2005
Taos Region

CHAPTER FOUR

Valle Vidal, Sangre de Cristo Mountains, Carson National Forest.

DECISION-MAKERS FIELD GUIDE 2005

Sustainable Development and Mining Communities

Dirk van Zyl, *Mackay School of Earth Sciences and Engineering University of Nevada*

The term *sustainable development* has been around since the early 1970s. However, it became firmly established in 1987 as a result of the report of the United Nations Committee on Development and Environment, where the concept was described as follows:

Humanity has the ability to make development sustainable-to ensure that it meets the needs of the present without compromising the ability of future generations to meet their own needs.

The accepted pillars or elements of sustainable development are economic, environmental, and social/community. These often are used with a modifier, such as economic equity, environmental well being, and social/community well being. Other terms that have been used to provide an image of sustainable development are the three-legged stool and triple bottom line. The former refers to maintaining a balance between the three components (the three legs of a stool), whereas the latter refers to an accounting sheet providing bottom lines (profit/loss) not only on economic activity but also on environmental and social/community well being.

Other aspects are also very important in defining sustainable development, including:

- Governance (of countries and companies)—often considered a fourth element or leg of the stool; the term governance refers to the laws and regulations of different levels of government as well as the capacity to implement and enforce those laws, however, it also refers to the policies, culture, and management of mining companies and their capacity to interpret and live by the applicable laws and regulations

- Technology—as technology changes so does our ability to change the contributions to technology

- Scale—sustainable development can have different meaning at the local, regional, national, and global scales

Ultimately sustainable development is a concept of needs, an idea of limitations, a future-oriented paradigm, and a process of change. In contributing to sustainable development, mining companies will have to consider the needs of communities, the limitations of their own resources, and often the limitations of the communities to participate in the technical discussions that are part of the permitting process. All of this must be done with a clear view of the future and will most definitely require changes in thinking and culture.

MINING AND SUSTAINABLE DEVELOPMENT

Society depends on many materials to maintain a specific standard of living or to improve its standard of living. There must be a sustainable supply of these materials to maintain economic activity and supply the needs of society. Mining is one way of supplying these materials; recycling and re-use are other options. However, for most materials the present international society is dependent on primary supplies from mining. At this time we cannot supply all the materials required by society from other sources without including mining.

Mines are developed where ore is found. Site, climate, topography, and other physical conditions determine the potential for positive or negative environmental impacts from the mine during the mining life cycle.

Mining and sustainable development does not refer to sustainability of the industry, a company, or a mine; sustainable mining is clearly an oxymoron, as all ore bodies will be depleted over time. However, the concept as applied to mining refers to a culture that addresses in very clear and practical terms how mining can contribute to sustainable development. Supplying the materials that society needs is one contribution, protecting or improving the well being of the environment is another, providing for the long-term well being of communities is yet another.

THE MINING LIFE CYCLE

Every mine follows the same life-cycle stages. However, the site-specific characteristics are different. The major life-cycle stages are exploration, mine development, operations, closure, and post-closure. The mine may also close temporarily because of low commodity prices, labor disputes, etc. The mine may also re-open when new technologies or higher metal prices make it possible again to have a profitable operation. Future land use for the mine site may include economic activities other than mining, such as solid waste disposal in the mine pit, construction of homes on waste rock disposal facilities or other mine property, renewable energy development, etc.

The timelines associated with each of the stages of the mine life cycle vary from site to site. There are mines with operating stages as short as three to five years, whereas others have operating lives of more than one hundred years. The mine life cycle is not a linear process; many things can happen during the various stages that change the outcomes in terms of longevity, environmental impacts, and future land use. It may be more appropriate to refer to the "mine life spiral." The intent of the "mine life spiral" is to improve the environmental and social/community well being through the development of a mine.

Past mining activities have left many legacies. These legacies include positive aspects such as long-term economic development of a region, as well as negative aspects such as environment degradation and communities subject to boom and bust economic activity. Negative legacies can jeopardize the development of new mines, and it is essential that these past legacies be addressed. A coordinated effort is required to accomplish the reclamation of mine sites, in addressing both safety concerns and negative environmental impacts. A number of agencies in the U.S. are active in the remediation of previous mine sites. However, more funding and coordination is necessary. Such a coordinated effort has been established in Canada between the federal and provincial governments.

There has been a remarkable evolution in the approach of mining companies to the mining life cycle. Until the 1960s it was common to leave a mine site as is when the project stopped returning a profit. The actions at that time did not include closure activities such as reclamation and site remediation. In many cases these mines were abandoned, and the associated communities became ghost towns. In the 1980s mining companies actively implemented reclamation at mine sites, and the philosophy of designing and operating for closure took hold in the 1990s. Much attention is currently paid to this stage of the mining life cycle. Extensive regulatory requirements are also in place in the U.S. for the closure of mines.

Examples of mines that have or are contributing to the sustainable development of communities and ecosystems are becoming more common. A recent paper by Kennecott Minerals describes the positive contributions of two projects: the Flambeau project near Ladysmith, Wisconsin, and the Ridgeway project near Columbia, South Carolina.

THE SEVEN QUESTIONS OF SUSTAINABILITY

One of the activities of the Mining, Minerals and Sustainable Development Project in North America was to develop an approach to measure the contributions that mining makes to sustainable development. The outcome of this project, with contributions from about forty individuals with widely varying backgrounds and interests from the U.S. and Canada, was a set of seven questions that can be asked about any mining (or other development) project. The questions must be customized to fit the specific interests and needs of a local community. So, although the questions may be universal in their application, they very much relate to the needs and interests of each community. The seven questions are:

> 1—*Engagement.* Are engagement processes in place and working effectively? The term engagement is used to describe a process of active listening and participation in discussions by both the mining company and the communities that are impacted.

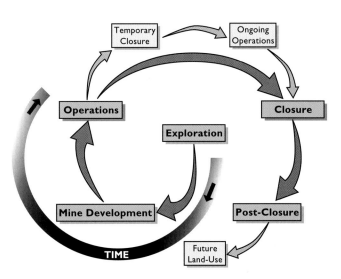

Mine life cycle stages. Future land use can include further exploration and ongoing mining, other economic uses such as renewable energy or grazing, and wildlife habitat.

2—*People (human well being).* Will people's well being be maintained or improved during and after the project or operation?

3—*Environment (ecological well being).* Will the integrity of the environment be taken care of in the long term?

4—*Economy (market economy).* Is the economic viability of the company assured; is the community and regional economy better off not only during operation but into post-closure?

5—*Traditional and Non-Market Activities (non-market economy).* Is the viability of traditional and non-market activities in the community and surrounding area maintained or improved with the project or operation?

6—*Institutional Arrangements and Governance.* Are the rules, incentives, and capacities in place now and as long as required to address project or operational consequences?

7—*Synthesis and Continuous Learning (Continuous Learning and Adaptive Management).* Does a synthesis show the project to be net positive or negative for people and ecosystems; is the system in place to repeat the assessment from time to time? A synthesis is required to combine answers from all six previous questions in making the determination of the contributions of the project to the well being of both people and ecosystem.

Methodologies have been proposed to implement these seven questions at the local level, and they provide a very powerful platform for evaluating the contributions of single or multiple projects to the sustainability of a community or an area.

An important realization during the development of the Seven Questions project was that a mine provides a bridge between the pre-mining and post-mining environment (nature and humans). This creates a significant opportunity to consider how the resources from the mining and other economic activity can be used to maintain or improve the well being of the environment and communities (in other words, how the mine can contribute to sustainable development).

Many mining projects contribute much to the communities where they are located. These contributions take place during the operating life of the mine, and it is challenging to find ways to expand these contributions so that future generations can also benefit from the mining activity. Active engagement of communities and mining companies is essential for this to occur. This can only be done if there is trust and respect between the different groups. Communities may not have the capacity to perform the long-term planning that is necessary to develop a sustainable development plan. Governments and companies must become partners of the communities to accomplish this.

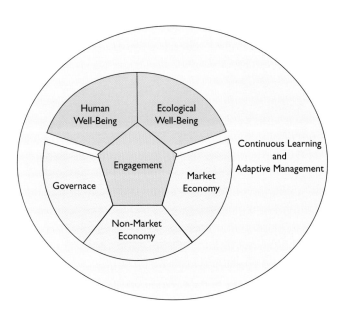

The schematic of the seven questions in this figure clearly emphasizes the centrality of engagement, ecological (environmental) well-being, and human-well being to the contributions of mining to sustainable development. Much of the focus must be on these issues. Schematic of Seven Questions (after Ian Thomson, On Common Ground Consultant, Vancouver, B.C.)

HURDLES TO IMPLEMENTATION

There are hurdles that must be overcome to maximize the contributions of mining to sustainable development. Our laws and regulations have not been developed with sustainable development as a foundation, and they may contain aspects that make it difficult to fully implement the laws and to contribute to sustainable development. One such hurdle is related to regulations that govern federal land use and post-mining economic activities on the land. It could be very beneficial for some communities to have full access to the facilities at a mine site so that they can be used for other economic activities—e.g., renewable energy, engine rebuilding, etc.

Another hurdle is the capacity of communities to fully implement sustainable development concepts and the associated activities. Coordination between companies, government, and educational institutions is necessary to keep that process going. Resources will have to be available to coordinate all these efforts. Such resources will have to come from a number of sources, including federal and state funding, private enterprises, and the mining companies. It is clearly unrealistic to expect that mining companies should carry the full load for such support.

A third hurdle is related to mining legacy issues. These are not always clearly understood or appreciated by industry, government, and civil society. Although coordination between federal, state, and local governments can do much to correct safety and environmental issues, mining companies must also understand the social legacy issues. These are not always the same for all communities; it is only with active engagement that they can be identified and addressed.

Mining contributes significantly to sustainable development of societies. However, ongoing efforts are required to expand these contributions and make them work in the long term.

Suggested Reading

Our Common Future: The World Commission on Environment and Development. Oxford University Press, 1987.

www.abandoned-mines.org This Web site provides information about the activities in Canada to address the issues related to abandoned mines on a national scale.

Breaking New Ground: Mining, Mineral and Sustainable Development. International Institute for Environment and Development, London, UK, 2002 (www.iied.org/mmsd).

Seven Questions to Sustainability: How to Assess the Contribution of Mining and Minerals Activities. International Institute of Sustainable Development, Winnipeg, Manitoba, 2002 (www.iisd.org/mmsd).

Fox, F.D., 2003, Mining and Sustainable Development: Flambeau and Ridgeway Mines—Lessons Learned; Kennecott Minerals Company, Salt Lake City, Utah, Paper presented at the Society of Mining, Metallurgy and Exploration Annual Meeting, Cincinnati, OH.

Shields, Deborah J. and Solar, Slavko V., 2000, Challenges to Sustainable Development in the Mining Sector, UNEP Industry and Environment, Volume 23, Special Issue, Mining and Sustainable Development II—Challenges and Perspectives.

Sustainable Aggregate Resource Management

William H. Langer, *U.S. Geological Survey*

The two main sources of aggregate, an essential commodity in our modern world, are crushed stone and sand and gravel. Buildings, roads, highways, bridges, railroads, airports, seaports, water and waste treatment facilities, and energy generation facilities all require large amounts of aggregate. Aggregate in one form or another is also used in many industrial, agricultural, pharmaceutical, and environmental applications.

About 16.8 million tons of aggregate worth about $94.5 million was produced in New Mexico in 2002. Sand and gravel was about 65 percent of the total; crushed stone was about 35 percent of the total. Although there are increases and decreases in annual aggregate production, during the last twenty-five years aggregate production in New Mexico has increased by about 73 percent. During the same period of time population has increased from about 1 million to about 1.8 million. Per capita consumption of aggregate has decreased slightly from about 9.5 tons per year to about 9.4 tons per year. New Mexico's population is expected to reach nearly 2.45 million by 2020. Based on those predictions, New Mexico will consume over 330 million tons of aggregate between 2005 and 2020.

Geology determines the location of an aggregate resource; location and inherent physical and chemical properties are non-negotiable. Although sources of aggregate are widely distributed throughout the world, there are large regions where aggregate is non-existent. Even if sources of aggregate are present, they must meet certain quality parameters before they can be used. There are large regions where chemical or physical properties of local aggregate render them useless for many applications. New Mexico is fortunate in having quality aggregate deposits scattered throughout the state.

Potential environmental impacts associated with aggregate extraction include the conversion of land use, changes to the landscape, loss of habitat, noise, dust, vibrations, erosion, and sedimentation, as well as impacts from the truck traffic that normally accompanies aggregate operations. Most of the environmental impacts associated with aggregate mining are relatively benign. By employing best management practices,

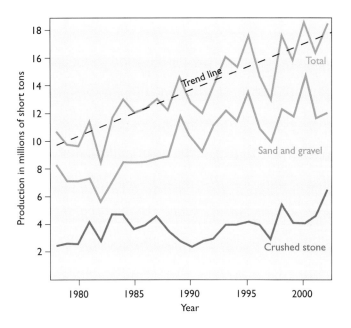

New Mexico aggregate production, 1978–2002.

most environmental impacts can be controlled, mitigated, or kept at tolerable levels and can be restricted to the immediate vicinity of the aggregate operation. Some otherwise high-quality aggregate resources may not be developed because of environmental reasons.

Aggregate is a high-bulk, low unit-value commodity that derives much of its value from being near markets. Thus, it is said to have a high "place value." Transporting aggregate long distances can add significantly to the overall price of the product, and in some situations can be the major component of the final delivered cost. Therefore, many aggregate operations are located near population centers and other market areas.

This juxtaposition of aggregate operations and population centers may result in conflicting land uses that prevent development of otherwise suitable resources. "Resource sterilization" occurs when the development of a resource is precluded by another existing land use. For example, aggregate commonly cannot be extracted from underneath a housing development or shopping center.

Despite society's dependence on aggregate, citizens may demand that crushed stone and sand and gravel not be mined nearby. To protect citizens from the impacts of mining, governments may require permits or impose regulations to control aggregate development, thus further restricting aggregate availability. This type of conflict, referred to as the "Dispersed Benefits Riddle," occurs because the negative impacts from aggregate resource development are usually located near the site of extraction, whereas the benefits from resource extraction are dispersed throughout an entire region. These dispersed benefits are not commonly considered in the local permitting process. But when aggregate resource extraction is denied because of opposition by the local community, other costs arise. Longer haul routes result in more traffic, more accidents, more fuel consumption, generation of more greenhouse gases, greater wear and tear on vehicles, and higher vehicle replacement rates. In such cases, gains by the local community may come at the expense of the greater public, the greater environment, and some other local area where extraction ultimately takes place. Instead of reducing impacts, they may simply be exacerbated and transferred elsewhere.

The reality is that every land use decision has both costs and benefits. The answer to the riddle is finding a means to ensure that the dispersed benefits of aggregate use are adequately weighed in resource development decisions. Sustainable aggregate resource management (SARM) might be the solution.

WHAT IS SUSTAINABLE RESOURCE MANAGEMENT?

The term sustainability dates back to the 1980 World Conservation Strategy, and was given prominence in *Our Common Future* (1987), otherwise known as the "Brundtland Commission Report." That report states that the purpose of sustainable development is to ensure that development meets the needs of the present without compromising the ability of future generations to meet their own needs. At a minimum, sustainable development must not endanger the natural systems that support life on our planet: the atmosphere, the waters, and the soils.

In the simplest sense, the "manufactured capital" and "natural capital" (natural resources) that one generation passes on to the next must be maintained or enhanced in order to achieve sustainable development. This philosophy gets somewhat confusing when dealing with a non-renewable resource such as natural aggregate (in contrast to a renewable resource such as forestry products). Aggregate resources, like all non-renewable resources, are indeed finite, and society technically cannot pass on the same amount of aggregate to its progeny.

However, unlike many non-renewable resources, the potential supply of aggregate resources on a worldwide scale is so large that "finite" has no practical meaning. Consequently, on a world scale, there is no real concern about running out of aggregate resources, and sustainability does not need to be invoked to ensure adequate future supplies of aggregate. But natural aggregate of suitable quality for an intended use can be in short or non-existent supply on a regional or local scale, and in the realm of sustainability, having an accessible local supply of aggregate resources takes on great significance because, as described above, transporting aggregate long distances adds to the overall cost of the product and to the overall cost to the environment.

An equally important goal of sustainability is to protect or enhance the environment. The aggregate industry manages a large amount of land and can promote sustainability by employing operational practices that minimize adverse effects on the environment and maximize the benefits from reclamation.

Sustainable aggregate resource management requires developing aggregate resources in an environmentally responsible manner that does not result in long-term environmental harm, even if short-term environmental impacts are unavoidable. There are many regulatory and voluntary tools that can be used to identify, reduce, and control negative environmental impacts, including best management practices, environmental impact assessments, environmental management systems, environmental accounting and reporting, and ISO 14000 standards. These tools can be applied both on site (quarry and processing facility) and to transportation routes.

SARM, however, is not just about protecting the environment from the potential negative impacts of aggregate extraction. Reclaiming aggregate operations or abandoned sites has tremendous potential to improve our quality of life, create additional wealth, restore the environment, and increase biodiversity. In today's expanding suburban areas, mined-out aggregate pits and quarries are converted into second uses that range from home sites to wildlife refuges, from golf courses to watercourses, and from botanical gardens to natural wetlands. Reclamation can be a major element of sustaining the environment and creating biodiversity.

Topographic map showing approximate location of gravel extraction near Sedalia, Colorado. The photo shows this same area following reclamation.

HOW DO WE APPLY SUSTAINABLE AGGREGATE RESOURCE MANAGEMENT?

To be effective, SARM must be a pragmatic pursuit, not an ideological exercise. It is an ongoing process among stakeholders, so government, citizens, and industry should all be involved in the pursuit. The process consists of a number of steps, including issuance of *policy statements*, elaboration of *objectives*, establishment of *actions*, identification of *indicators*, and *monitoring*.

Policy statements issued by governments commonly identify the aggregate industry as a key industry contributing to jobs, wealth, and a high quality of life for its citizens, and commit the government to the protection of critical aggregate resources and the protection of citizens from unwanted impacts of aggregate extraction. Industry policy statements commonly identify environmental and societal concerns and commit the company to environmental stewardship and interaction with the community. *Objectives* describe what is intended to be accomplished. *Actions* are associated with each objective and describe the approach needed to reach the objective. Examples of paired objectives (in italics) and actions include:

- *Maximize availability of, and access to, aggregate* by forward planning that protects important resources from urban encroachment; by extracting as much aggregate as possible from an area that has to be disturbed and using it for the most economically valuable application appropriate for the aggregate quality; by avoiding high grading (picking the best parts of the resource and limiting the ability to use the remainder); by finding uses and markets for all of the material disturbed (e.g., turning crusher fines into "manufactured sand" thus taking the pressure off natural sand sources in more environmentally sensitive areas); and by encouraging use of recycled aggregate.

- *Minimize societal impacts* by forward planning that protects communities from the nuisance impacts of poorly designed, poorly located, and poorly managed aggregate operations; by using best management practice designs and operations to control blasting, noise, dust, sediment erosion, and visual scarring in extractive and transport operations; and by involving the local community in planning activities. Community involvement may lead to a measure of community acceptance and an unofficial "social license to operate," which can be just as important as the official, legal permits.

- *Minimize environmental impacts* by providing for conservation of natural surroundings with buffer areas that maintain or enhance vegetation and wildlife habitats and corridors; by using best management practice designs and operations.

- *Maximize rehabilitation of disturbed areas* by allowing for reclamation as an integral part of the quarry/pit design process before extraction begins; by starting rehabilitation from day one; and by being flexible enough to allow for advances in technology and changing local needs.

MONITORING SUCCESS

Indicators measure progress toward reaching objectives as well as the effects of actions to protect and enhance natural and human systems. Indicators are specific to the target and actions but tend to be similar in many situations. Example indicators include:

- Proportion of aggregate coming from areas preferred for extraction

- Proportion of natural aggregates compared to recycled material

- Proportion of sites covered by modern operating conditions

- Proportion of aggregate coming from environmentally sensitive areas

- Area of land restored compared to area of land undergoing extraction

Monitoring, feedback, and the regular reconsideration of requirements as events develop all help to refine the SARM process. Establishment of a joint monitoring process presents an excellent opportunity to forge partnerships with communities and involve citizen groups.

To ensure the sustainability of aggregate resources, each of the primary stakeholders must accept certain responsibilities:

- The government is responsible for developing the policies, regulatory framework, and economic climate that provide conditions for success.

- The industry must work to be recognized as a responsible corporate and environmental member of the community.

- The public and non-governmental organizations have the responsibility to become informed about natural resource management issues and to contribute constructively to a decision-making process that addresses not only their own, but also a wide range of objectives and interests.

- All stakeholders have the responsibility to identify and resolve legitimate concerns; government, industry, and the public must cooperate at regional and local levels in planning for sustainable aggregate extraction.

Sustainable aggregate resource management, and finding an answer to the "Dispersed Benefit Riddle," would be less difficult if all conflicts between regional aggregate resource needs and local impacts had solutions that would leave everyone better off. This is seldom the case, and there are usually winners and losers. But as the amount of accessible land that is underlain with suitable aggregate resources diminishes, inequalities will increase. The longer we wait to implement sustainable resource management principles, the more difficult it becomes to implement sustainable resource management.

Surface and Ground Water Management Practices at Mining Operations

R. David Williams, *Bureau of Land Management*

Many of the challenges facing the minerals industry in the twenty-first century, a century that must focus on sustainable development, center on the development of technologies and practices that can help to eliminate or limit long-term environmental liabilities during mining and following closure of mining and processing facilities. Often the issues surrounding mine operations and closure focus on the control, management, and treatment of surface and ground waters, and on the disruption of ground and surface water flow paths during excavation of large surface mine open pits. Most agencies and companies try to follow a fairly standard set of procedures to minimize long-term environmental risks to ground and surface waters, including *avoidance, isolation,* and *treatment.*

Avoidance can be accomplished either by not disturbing or exposing problematic waste units (i.e., tailings and waste rock piles) encountered during mining operations, or by avoiding specific mineral processing technologies that may create these problems. These choices affect the project economics, so in many cases avoidance is simply not a practical alternative. In this case, the next alternative is to isolate problematic waste from air and water that can create and transport contaminants off site. *Isolation* can incorporate several alternative strategies and technologies. Problematic waste can sometimes be minimized through selective mining or handling, or operational changes in milling processes. Waste can sometimes be treated in place through addition of amendments. Isolation by treatment in place is an attempt to avoid the final alternative: *treatment.* Long-term treatment is the least desirable alternative, because it generally requires a continued infrastructure for treatment in addition to the treatment itself, and is generally the most expensive alternative.

SURFACE WATER CONTROL AND MANAGEMENT

Surface water control can be an important aspect of a sustainable development program. A poorly designed surface water control program will result in long-term maintenance issues, and may result in additional waters needing treatment, or in violations of water quality standards. Because modern mine sites are typically large-scale disturbances of the existing environment, drainages and streams may be disturbed or eliminated, and replaced with constructed drainageways that must be designed to accommodate the large storm flows that will occur intermittently for many years to come. Current mine reclamation practices tend to include the construction of extensive areas of uniform slopes, broken by benches and drainageways that effectively transport rainfall and snowmelt off reclaimed slopes. However, they do not mimic the natural environment. They are essentially engineered terrains and, like all engineered features, require routine periodic maintenance to remove debris and repair damage from unanticipated events.

This catchment pond at the Ortiz mine in northern New Mexico was built to prevent mine drainage from entering surface streams.

A developing technology involves the shaping of waste dumps using drainage patterns and slope forms that mimic the surrounding terrain, typically including complex slopes and dendritic drainage patterns. Global positioning system technology is very helpful in the application of this new reclamation technique because it allows earth-moving equipment to achieve far greater accuracy when creating specific slope angles and elevations. BHP Billiton is currently using this technique at coal mines in New Mexico, and the Montana Department of Environmental Quality and the federal Bureau of Land Management are actively

Modern mine reclamation often includes extensive areas of uniform slopes, with slope breaks/benches and constructed drainageways.

encouraging its use in future mining proposals in Montana. The goal is to help limit long-term maintenance requirements by developing reclaimed environments that more closely mimic the appearance and function of the natural environment. This technique has been used extensively on many smaller-scale abandoned mine reclamation projects.

GROUND WATER CONTROL AND MANAGEMENT

The first step in a ground water control program is an effective surface water control program, because the two are related. A ground water control program for reclamation generally is distinct and separate from the ground water control program for active mining. Active mining programs focus on removing water from the mine in order to conduct operations in a relatively dry environment. Following the completion of mining, the program objectives will likely shift to either avoid ground water interaction with problematic waste material, or to collect and control contaminated waters. Ground water control is an art unto itself, and techniques vary from site to site. Ground water controls at mine sites can include sheet piles, grout curtains and clay cutoff walls that seek to isolate a project area from outside ground water. However, most mines rely on well systems that intercept the ground water and pump it away from the project area. The other technologies mentioned tend to be much more complex and are generally better suited to smaller-scale projects.

As most mine pits are below the water table, the ground water will reestablish its approximate original elevation in the area once pumping stops. If the pit is not backfilled, a pit lake will result. Pit lakes have been proposed at many mine sites. They may be an acceptable closure strategy if the pit lake water quality issues can be satisfactorily resolved. There is considerable ongoing research into the use of pit lakes as water treatment facilities relying on in-pit processes to treat the water. Research has focused on how mixing and stratification of pit waters can be used to enhance or limit chemical processes to improve water quality.

Although it is appealing to think that backfilling pits with mine waste will resolve environmental issues, success depends on site-specific conditions. If the waste does not have the potential to become acidic and does not contain contaminants, this can be an excellent reclamation strategy that eliminates the waste rock and the pit. However, if the pit is backfilled with acidic mine waste, ground water may interact with the backfilled material, forming contaminants

Reclamation and slope stabilization on tailings at the Miami mine in Arizona. The cows are grazing on the newly reclaimed slope to the right.

that will adversely impact water quality. In this case, the resulting poor-quality water would have to be collected and treated or isolated in place, and both of these options are expensive and difficult to achieve. Therefore, not backfilling the pit and dealing with the resulting pit lake may be less costly and have fewer ground water contamination problems in the long run.

TREATMENT OF MINE IMPACTED WATERS

Mine impacted waters (MIW) may need to be treated in order to meet water quality standards. This may be done as a last resort after other control methods have proven inadequate, or generation of MIW may be an unavoidable impact of mining even though proper surface and ground water control programs can help to limit the amount of MIW needing treatment. The types of treatment for MIW include (in order of relative cost) passive, semi-passive, and active. Passive treatment technologies rely on natural chemical or biological processes to precipitate or remove metals and other contaminants from the MIW. They do not require continual intervention, though they do require ongoing monitoring and maintenance. Typical passive systems include constructed aerobic and anaerobic wetlands, anoxic limestone drains, vertical flow systems, and open limestone channels. Active treatment technologies require a regular staff and infrastructure, relying on a continuous source of power and materials to maintain the treatment system. Active treatment is generally considered to be a safe and effective technology and operates at a large number of mine sites. The most common types of active treatment include chemical or biological addition, filtration, and ion exchange. Semi-passive treatment often consists of an active treatment with a passive treatment as a finishing step in order to meet water quality standards.

The most common form of treatment for MIW is chemical addition of alkali, usually limestone, to raise the pH (lower the acidity) of the solution and precipitate the metals out of solution. Filtration is generally used for the removal of particulates, though the filtering media can be used for some limited chemical reactions if concentrations are low. Ion exchange is also most suitable for relatively low concentrations of metals. A variety of natural and artificial products can be used to remove metals from MIW through cation exchange. As with the other processes noted above, the materials used must be replenished as they are consumed. Such treatment may be necessary for many years or in perpetuity. Active processes also result in

Pit lake at the Ortiz mine in northern New Mexico.

the production of a waste product, generally a sludge that must be disposed of appropriately. Because of the expense associated with active treatment, there is considerable research focusing on new technologies that reduce treatment costs. Much of this research involves the use of biological processes to reduce either the cost or quantity of reagents and materials needed.

New and Evolving Technology in Mining and Mineral Processing

Douglas Bland
New Mexico Bureau of Geology and Mineral Resources

Mining companies are continually looking for innovative ways to reduce operational costs, increase mineral recovery, improve working conditions, and minimize negative impacts on the public and the environment. Research in the pursuit of new technologies that help to achieve this goal is ongoing. Large corporations have their own research and development (R&D) programs, and much of the work underway in this setting is proprietary and may not be available to the public.

Other R&D efforts are undertaken by companies that provide specific products or services. One example of this is a new technology using microbes to recover metals from acid mine waters. BioteQ Environmental Technologies, Inc., a Canadian industrial process technology company, has developed and the patented BioSulphide Process® for treatment of acid mine drainage that is being used by the Phelps Dodge Mining Company in Bisbee, Arizona, for the recovery of copper. BioteQ's process is relatively inexpensive compared to conventional ion exchange processes, allowing economic recovery of marketable metals from low-grade solutions.

At Bisbee, acid drainage is collected at a water plant and pumped to the BioteQ facility. A sulfur-reducing anaerobic bacteria culture is used in a bioreactor where it plays a key role in producing hydrogen sulfide gas. The acidic drainage, which contains dissolved copper and other metals, is mixed with the hydrogen sulfide gas. Copper then precipitates out as a sulfide concentrate and is shipped to a smelter for further refining. The process water is pumped back to the stockpile where it is re-applied to continue leaching copper. The bacteria never enter the water and are contained in the bioreactor. Other metals, such as nickel, cadmium, zinc, and iron, can also be precipitated from acid mine drainage by adjusting the pH to near neutral.

Another example was developed by an Australian company, Virotec International, Ltd., that created a series of reagents or substances that alter the chemical nature of problematic materials. These reagents are applied at mine sites in powder, pellet, or liquid slurry form directly onto existing acid-generating rock and leach piles, acid tailings ponds, other contaminated waters, and soils where they neutralize acid materials and remove excess metals without generating toxic waste products. These reagents have been used successfully at a number of mines around the world.

Other R&D programs are partnerships between mining or service companies, the government, and research facilities. For example, the U.S. Department of Energy's Office of Energy Efficiency and Renewable Energy manages the Industrial Technologies Program (ITP), whose mission is to improve industrial energy efficiency and environmental performance through projects involving partnerships with industry, national laboratories and other research institutions, and stakeholders.

One project the ITP saw through to fruition was implemented at the smelter associated with the Bingham Canyon copper mine near Salt Lake City, Utah. Kennecott recently installed new burners in the smelter that use a new technology called dilute oxygen combustion, which incorporates high velocity fuel and oxygen jets that produce increased combustion in the furnace. Use of this technology has increased efficiency, reduced fuel needs, and reduced nitrogen air contaminants (NOx) by 80%.

Other ITP projects that have completed their R&D or are underway include:

- Robotics for improving mining productivity
- Projectile-based excavation
- Treatment of cyanide solutions and slurries using air-sparged hydrocyclone
- Robot-human control interactions in mining operations
- Development of new geophysical techniques for mineral exploration and mineral discrimination based on electromagnetic methods
- Investigation of a combined GPS and inertial measurement unit positioning system for mining equipment
- Imaging ahead of mining
- Drilling and blasting optimization
- Mining byproduct recovery

There is no guarantee that all of these projects ultimately will achieve success and be incorporated into industry practices, but many others will be investigated and adopted, changing the way mining and the environment are managed in the future.

How Science Can Aid in the Decision-Making Process—Translating Science into Legislation

Peter A. Scholle, *New Mexico Bureau of Geology and Mineral Resources*

Science, like most fields of inquiry, represents a search for complete understanding or "truth." But scientists are no more or less infallible than others in our society. They can be unintentionally biased; they can be (one hopes only in very rare instances) intentionally biased (remember the American Tobacco Institute?); they often work with incomplete data sets making it difficult to get reliable conclusions; and they can draw incorrect conclusions even from fairly complete data sets. But the process of science is one of constant questioning and testing of one's own work and the work of others. That jousting and sparring between scientists, no less than the competition between animals in the wild, leads to the survival of the fittest—the fittest ideas and conclusions in the case of scientists. The process of intellectual competition can be a long and slow one. Indeed, it is really a never-ending process as new data are gained, and old ideas are reexamined. Thus, scientific inquiry is basically a process of successive approximation in which we draw ever closer to the truth, but perhaps never completely reach it. That is no less true in economics, history, political science, or any other field of inquiry (think of polling results both before and after the last election, for example), but common public expectations are that scientific inquiry should rise above such data uncertainties to yield definitive answers.

For decision makers in particular, dealing with scientific uncertainty can be a very frustrating and even disillusioning process, especially as they listen to scientists debate their individual perspectives on technical issues that may have significant social implications. That frustration, in turn, often results in legislative gridlock, in calls for more research to resolve the conflicts (thereby sometimes providing political cover for inaction on the real problem), or in the waste of funds expended for actions that later prove to be ineffective or useless.

The global warming/climate change issue provides a good example of such controversy within the scientific communities and the struggles of decision makers throughout the world to deal with the associated scien-

tific uncertainties. It may help us to understand the differences between measurable, predictable, or modelable information in an exceedingly complex system. Let's start with the simplest question—is the earth's surface warming? Simple—we just stick the thermometer in ... where? Well, even that is complex. It is a big planet, and we have a limited number of accurate ground-based weather stations, most of which are located in North America and Europe. So we send up expensive satellites that scan the earth as they orbit around it, and we compile global temperature information. But how do we compare that modern information with ground-based temperature data from 100 years ago to see when warming began and how far it may have progressed? This remains an area of conflicting opinion among scientists, but an overwhelming majority of reputable (and who decides that?) scientists now agree—warming of the order of 1° C has occurred over the past 100 years. But that just takes us on to the harder questions with less easily measurable answers: have we caused such warming through human industrial and agricultural activities (deforestation and burning of fossil fuels, for example), or is it part of a natural climate cyclicity that geologists have deduced to have occurred over the past billions of years of earth history? If we caused the changes, how can we model how quickly warming is proceeding and predict what the consequences will be? Are other things we are doing (pumping sulfur dioxide and particulates into the atmosphere, among others) offsetting all or part of the effects of greenhouse gases?

Clearly, the answers to these questions are difficult to obtain and may never be known with full certainty. The United States has responded to the crisis with a call for more research before taking any action. Much of the rest of the world has responded with a desire to immediately start to reduce greenhouse gas emissions. These completely different legislative responses to the same data and interpretive models highlight the difficulties of dealing with uncertainty—we will know the right answer in 100 to 200 years, but by then it may be too late to change our choices.

So how can science aid in the decision-making process? What can decision makers do to deal with such uncertainties? The first step is to stop thinking just about right and wrong and think instead about probabilities. There is virtually nothing we ever know with 100% certainty. We all make decisions in our private lives based on probability. You probably do not believe you are going to get sick or die this year, yet you most likely do have medical and life insurance.

Why?—Because the consequences of being wrong are significant enough and the possibility of getting ill or dying is great enough that you take prudent steps to deal with the problem. The same insurance-oriented, probability-based, decision-making principles should be applied to water, climate, pollution, and other issues that have potentially drastic societal consequences.

A second step is for decision makers to take a more direct role in some aspects of science management. I am not advocating that legislators, for example, decide how fundamental science is done or interfere with the generally impartial peer review of technical proposals that forms the foundation of how science is done in most developed nations. But at the same time, it is unrealistic to expect that fundamental societal questions will be adequately addressed in the absence of guidance from the people who have to make the final decisions. Legislators and legislative staffers need to educate themselves on issues sufficiently to pose the questions that they feel need to be answered (in part by attending conferences such as this one). Then they need to ensure that funding is available to address those questions. Where funding is provided through agencies that put out topically specific requests for proposals (RFPs), rather than passively reviewing independently conceived and submitted proposals, research can be directed to targeted goals. Again, I want to emphasize that there is great value to independently conceived research, and we should continue to support it as well. But if you want answers for your questions, you should frame the questions and fund the search for answers directly.

In that same context, where legislators are authorizing and funding recurring large-scale, but still experimental work that has substantial scientific uncertainties—something like salt cedar eradication, for example—why not include a requirement (and funding) for independent scientific monitoring at the same time? That would assure that adequate data are available to judge the efficacy of such programs when subsequent requests come before the legislature.

Three other methodologies also can be recommended to deal with uncertainties and help translate scientific knowledge into better legislation. Two of these approaches shift the responsibilities of evaluating conflicting scientific opinions to other, potentially more knowledgeable groups. At the federal level, the National Academy of Sciences (NAS) acts as an organization that sets up specific, theoretically impartial review boards consisting of nationally known scientific and technical experts to review questions of complexity and importance (such as global warming). Although the process is not without its flaws, the consensus reports of these review boards carry considerable weight and commonly offer excellent guidance to legislators and the public. The effectiveness of this process depends substantially on how the reports are funded, how the panels are selected and chaired, and how the study questions are posed. The NAS will work, in some cases, on state problems, especially if the problems carry over into a number of states (that really means they prefer to work on regional or national issues). However, NAS reports are expensive; generally too expensive for individual states.

Independent, non-regulatory state agencies—people like the New Mexico Bureau of Geology and Mineral Resources—can sometimes act in a manner comparable to the NAS in order to deal with complex issues. The agencies can assemble panels of capable scientists, chair the meetings, and take on the responsibility of producing the final report authored by the full and diverse panel. It is not, however, realistic to expect that any state agency will take on the role of sole arbiter of complex technical and scientific questions. No single agency has the breadth of in-house expertise and perspective to compete with a broadly chosen board, and no state agency will eagerly (or should) take on the responsibility and potential liability of providing a definitive recommendation on complex issues of major social and financial import.

A final suggestion for the future of legislative decision making is the development of computer-based modeling tools specifically designed for legislators. Modeling tools, if properly designed, allow one to see the consequences of actions before they are taken. Well-designed models can also combine specific technical strategies with economic prognostication, allowing decision makers to understand the economic consequences of specific actions or the lack of any action. To go back to our climate change question: a well-designed model should allow one to predict the costs of doing nothing to curb greenhouse gas emissions (costs related to rising sea levels, changing ecosystems, higher temperatures, more extreme storms, and other effects) versus the costs of reducing emissions (costs such as building infrastructure for alternative energy sources, increasing efficiency of energy use, carbon dioxide sequestration programs, and others). Models can be graphic and effective tools for prediction, but models are no better that the programming and data that go into them. If the linkages and feedbacks in a system are not well understood, if data for one or more of the variables are not available or have large uncertainties, then the effectiveness of the model is greatly reduced. Small changes in climate models, for example, often produce large variations in their predictions. The world is complex, and in systems with many variables, the number of potential interactions between variables grows exponentially. Good examples of this are drug interactions in humans or the interactions of combinations of chemicals released into the environment. We need to test not only if a drug or chemical is safe and effective taken or used by itself, but we also must test how it might interact with the thousands of other drugs on the market or other chemicals already found in surface or ground waters.

We have learned much about the world around us, but there is, oh so much more to learn. In the mining area, as well as in most other fields of environmental interest, we know the basics and need much additional information before we can understand the details. But all this should inspire you to fund more scientific investigations, not to give up on science in decision making. Science will not provide the answer to what are basically social choices but it can inform those decisions. Scientific information will never be complete, but using scientific input, no matter how incomplete, is better than ignoring the existing body of scientific data. The final decisions involving major human issues are always based primarily on social and/or economic considerations, but scientific input has a valuable place in narrowing the debate to focus on the most acceptable choices.

Incorporating Science and Stakeholder Values in the Decision-Making Process

Andrew MacG. Robertson and Shannon Shaw
Robertson GeoConsultants, Inc.

Planning, constructing, operating, and reclaiming a mine in today's world involves many specialized disciplines, some technical and scientific, and others related to public needs and values. All these factors must be considered to make the best choices that meet the needs of the mining company, regulators, and the public, yet it is not feasible for decision makers themselves to become well versed in all of these disciplines.

One fundamental problem that has plagued everyone concerned with these issues is how to combine quantifiable scientific data and qualitative social desires in one unified evaluation process that identifies the best solution, considering all these factors and giving appropriate weights for each. With the push for sustainable development across the industrial sector, a number of tools are evolving in order to do just this. Two tools that are becoming more prevalent in the mining industry are the *Failure Modes and Effects Analysis* (FMEA) for risk evaluation, and the *Multiple Accounts Analysis* (MAA). The latter is used for evaluating the impacts of various mining and reclamation options, and then identifying the most desirable options that minimize negative impacts and risks and maximize benefits, while considering the level of uncertainty. Both tools provide a means by which experts can convey their science-based assessments of various impacts and risks to other decision makers and other interested parties.

Multiple Accounts Analysis allows decision makers to select the most suitable choice from a list of options by weighing the relative positive and negative impacts of each. Issues used to develop options are grouped into four categories called accounts: technical, environmental, socio-economic, and project economics. Each issue is defined by indicators, some of which are straightforward and quantitative (costs, for example). However, many indicators, particularly environmental and socio-economic factors, are difficult to describe or quantify. Expert judgment is needed to assign relative levels of desirability (values) for certain indicators, based on scientific testing, modeling and analyses, precedents, and experience. Having participants who are experienced with similar projects and dedicated to understanding and learning the benefits and limitations of certain indicators is critical to the success of these evaluations. For example, the issue of water quality impacts is nearly always included in the evaluations of mining projects. Although a great deal of science is involved, the predictive values for long-term water quality are often difficult to quantify, so the ultimate evaluation ends up being qualitative.

Once the list of issues (accounts) is complete, the options are scored through a numerical process, and a matrix is developed that identifies the most favorable option. (Use Robertson Figure 2a to illustrate?) The MAA format serves to translate detailed scientific evaluations into justifications for each option that can be readily explained to decision makers, who can then defend their decisions using the expert evaluations in the MAA matrix.

The *Failure Modes and Effects Analysis* (FMEA) is another tool being used widely in the mining industry for assessing the risks associated with a preferred option. Risk is a combination of the likelihood of an event or failure and the consequence if it occurs. Often the effects of a failure can have impacts of different severity with respect to the economy, the environment, land use or biota, health and safety of humans, and regulatory compliance issues. Descriptions of likelihood and consequences are usually quantitative (science based) but occasionally qualitative (experience based, from an appropriately experienced scientist).

The FMEA allows decision makers to perform a systematic and comprehensive evaluation of the potential failure modes of an option in order to identify potential hazards. For example, the FMEA can be used to evaluate the collapse of a collection pond dam (failure) causing a discharge of water. While the likelihood of this in a semi-arid environment may be considered *moderate* (0.1 to 1% chance of occurrence in any one year), the consequences on the environment may be different for different circumstances. Depending on the quality of the discharged water and the type of environment into which it flows, the consequence could be *moderate* (science based and quantitative) with respect to the biological and land use impacts; the regulatory and public image consequences could be expected to be *high* (qualitative); and the cost consequences (science-based cost calculation) to repair the pond could be relatively *low*. Typically the FMEA work sheet and results are both developed as a matrix.

There is a difference between the risk of a failure and uncertainty in the estimate of that risk. There are also uncertainties associated with the potential frequency of failure and expected consequences. Quantifying the uncertainties provides decision makers with an understanding of the analyst's opinion in terms they know and understand.

Finding Solutions—Collaborative Processes and Issues Resolution

Julia Hosford Barnes and Mary Uhl, *New Mexico Environment Department*

Given the complexities of environmental problems today, governors, state agencies, legislators, and scientists are looking for long-lasting solutions that navigate political minefields. The converging worlds of science, government, and politics pose fascinating and extraordinary challenges to environmental public policy issues. Scientific information is technical, complex, and difficult to translate into the language of the layperson. Health and environmental issues often fall within the jurisdiction of different federal, state, and local governing bodies, and the issues can be politically heated when the goals of the stakeholders are diametrically opposed. Success in finding consensus or concurrence in these converging worlds is a large achievement.

The mining industry in New Mexico faces all of these challenges today. Most mining issues are related to: (1) siting—should a mine be located here; (2) impact on the local community; and (3) environmental protection and reclamation after mine closure. Identifying solutions acceptable to all stakeholders is complex and time-consuming, but often avoids protracted and costly litigation later on.

To improve the chances of long-term resolution, leaders are using the power of assembly to resolve complex environmental issues in mining and other sectors. In a collaborative process, community groups, the environmental community, industry, and government officials come together to:

- Develop a common understanding of the factual and scientific issues presented
- Design a process that identifies the goals and how to reach these goals
- Discuss and come to consensus on key issues
- Determine the steps to implement the decisions made

The work is typically done with a facilitator who helps to manage the process and a project manager who understands the technical and scientific issues involved. This team assists the convener to work toward completion of the project. (The team of convener, facilitator, and project manager are referred to collectively as "convener" here.)

Through our work at the New Mexico Environment Department in convening community groups to address local environmental issues, we have identified four key leadership goals and four key science goals that can improve the chances of a positive outcome.

LEADERSHIP GOALS

When a leader uses the power to convene a community group to work on a local issue, there are four areas that warrant special attention:

1—Manage carefully the initial design of the collaborative process. Two points in a process provide times when decision making is key: at the beginning and the end of a process. First, how you are going to work together and, at the end, how you are going to implement the agreements after the process concludes? The convener has several important initial decisions to make. Among these decisions are:

- Is the leader or the lead organization neutral and viewed as neutral? If not, can they still convene the process?
- What parties are important to bring to the table?
- What is the goal of bringing the group together?
- What are you trying to achieve and why?
- Do others involved agree with your process design?
- What does a successful outcome look like?

When the process is fully designed, the convener should be able to answer *Who? What? Where? When? How?* and, most importantly, *Why?* Clear goals at the beginning of a process can set up good decision making at the end of the process.

2—Manage the expectations of all involved. It is important to articulate goals regarding the process—both *what* you want the group to accomplish (e.g., to provide input on a health risk assessment) and *why* this is important (e.g., to have a better overall picture of the affected area or to work with industry to make safety improvements). The process differs depending on the outcome you seek. If you do not have clear goals, you run the risk of having various groups expecting various outcomes, and inevitably some of these expectations will never be met. For example, if your goal is to make safety improvements in targeted industries, industry must understand and support the desired outcome. If your goal is to help the community understand the overall environmental picture in the affected area, you will focus more on uncovering all sources of pollution in the area. If community groups expect industry to change its safety requirements, but the agency goal is only to understand the environmental impacts in the area, the community groups will be frustrated with the ultimate outcome.

When these goals are determined, it is essential to communicate the goals to the stakeholders and to those watching the process so that all are clear on what to expect from the process. Some projects are high profile and may be covered by the media. Typically, the upper echelons of government and industry are watching the process. New leaders can come in during the process, through elections or otherwise. It is important to be able to communicate effectively the project goals to new leadership, and to get the support of the new leaders. Clear goals and clear articulation of these goals can ensure that everyone expects the same things from the process.

One area of common confusion is the extent of power a governmental agency has under its regulatory authority. Whereas regulatory agencies understand how they are limited by laws under which they operate, the public commonly does not. The convener should ensure that everyone understands the limitation of the laws and what constraints that puts on the process.

We urge leaders to consider one note of caution: Public policy collaboration is sometimes suggested because nothing else has worked, and the situation has hit "rock bottom." Collaborative processes in these situations often achieve only modest results and may not achieve anything at all. It is difficult in these situations to come to consensus regarding every step in the process, from who is at the table to what you are trying to achieve. If a group is convened over a long-term dispute, it is best to set small clear goals and take the project one step at a time. Expecting a ten-year dispute to be resolved because a task force is convened is typically an unreasonable and unachievable expectation.

3—Manage the bureaucracy. Bringing a group together can be a cumbersome process. When a governmental agency acts as a convener, it necessarily has to work within the bureaucratic structure of government. If not managed carefully, the group process can be substantially less efficient due to the constraints of governmental bureaucracy.

The convener should anticipate the bureaucratic difficulties and then concentrate as many of the bureaucratic requirements into one time period, if possible. For example, if several contractors are needed, the community group could meet to decide the scope of work for all contractors. The group could then recess during the time needed for the agency to go through the process of selecting contractors. Once the contracts are in place, the group can then meet again to move forward. Bureaucracies tend not to move faster when pushed, so it is best to plan around these requirements.

4—Manage the process "All the world loves a stage." Group collaborations, by definition, bring people together. If forums are not managed appropriately, it provides an opportunity for all types of grandstanding by divergent groups. Decision making in the spotlight can be explosive and ineffective. It is important to manage the process to minimize the abuse of the public forum.

At the same time, the public forum can bring about good results. Statements made and decisions announced in public can make people more accountable. Agreements made by groups in a public manner are frequently honored. The convener can work to set ground rules for conduct at the meetings and can capture the decisions of the group so that the process moves forward.

SCIENCE GOALS

When groups are convened to look at scientific information, special challenges arise. The convener should ensure that scientific information is conveyed in a manner that can be understood by the non-scientists who are at the table. We suggest four areas on which to focus attention:

1—Provide baseline information on the relevant science so there is a common understanding in the group. In order to ensure that the scientific presentations are comprehensible, it is important to discuss the baseline scientific information at the beginning of the project. At this time, you can ensure that all members of the group understand the scientific concepts involved, understand how science might answer the questions asked, and understand the possible limitations to the answers. The baseline scientific discussion can also identify areas in which there is conflicting scientific information or protocol. These are areas that can be discussed later.

2—Provide understandable scientific information presented by scientists who can clearly deliver the information. Environmental issues can require many complex scientific components to be examined and discussed within a public forum. Some group members are able to understand complex scientific issues, and some may not. Scientific information should be presented without scientific jargon following the suggestions below:

- Choose a charismatic speaker who likes to present to the public. The speaker must listen to questions and respond both to the precise question asked and must address the underlying, unasked question. If English is the second language of the presenter, make sure that language or the presenter's accent does not make communication more difficult.

- Present the results of the research *first*, rather than beginning with the basis for the results. The technical basis for the results is important in the scientific world, but will not be followed easily by a layperson.

- Put the results into context for the public. The public needs to understand *why* a result is important.

- Make the presentation to a technical group *before* making the presentation to the larger public group and make necessary changes. A peer review committee of scientists is recommended. This can substantially improve the presentation to the public.

3—Provide a reality check for the public. Scientific studies can be very expensive, and they sometimes provide limited and conflicting answers. Most members of the public do not understand or appreciate how complex it can be to determine the scientific answers to questions asked. The convener should ensure that the members of the group understand the work necessary to obtain the requested scientific information, the limitations of the budget, and the possibility of limited or inconclusive results.

4—Provide a common agreement within the group about what to do with the dilemmas of misinformation, conflicting information, and perceived information. The convener should discuss ground rules with the group regarding how to handle misinformation, conflicting information, and perceived misinformation. This can be a very difficult problem because groups in conflict tend to question the "facts" presented by the other side, and it is time consuming to research each questioned "fact." These problems can quickly derail a group. On the other hand, if misinformation is not addressed, the group might act upon bad information. One possible solution is to set aside a portion of the budget to investigate factual disputes or form a technical peer-review committee that will look into factual issues.

Collaboration can result in new ways of working on troubled issues that concludes with a solution that is more palatable to the local community. It can provide long-term solutions to problems that affect entire communities. Although the process may be challenging, it can provide results that surpass any alternative. There is a great deal of information available regarding public policy collaborations. One excellent resource established to support governmental public policy projects is the Policy Consensus Initiative (http://www.policyconsensus.org). This is important, challenging, and exciting work. We wish you luck.

The Role of Non-Government Organizations

Brian Shields, *Amigos Bravos*

The current NGO policy issue centers on creating an enabling environment for NGOs to play equal roles in the development of the society especially when the government is withdrawing from the social delivery processes. —Tanzania Gender Networking Programme

The World Bank defines non-government organizations (NGOs) as "private organizations that pursue activities to relieve suffering, promote the interests of the poor, protect the environment, provide basic social services, or undertake community development." Although the World Bank's definition of an NGO may be an accurate portrayal of public interest organizations, it does not take into account that there are now a number of corporate-backed NGOs, such as the New Mexico Mining Association, promoting industry's agenda. I will limit my comments to the role of the community-based, environmental and social justice NGO as exemplified by Amigos Bravos. In exploring the role the NGO plays as it becomes increasingly invested in a community, I will draw on my experience working with Amigos Bravos to address the contamination of the Red River from mine-generated waste at the Molycorp mine in Questa, New Mexico.

The primary role of the environmental and social justice NGO is to be a voice for the long-term health and well-being of individuals, communities, and the natural environment. It is the NGO's responsibility to help identify and define the problems and issues that impact the health of the community and its environment, to advocate for policies and actions that promote a healthy and sustainable existence, to hold government, industry, and polluters accountable for actions that are detrimental to a healthy life, and to help develop solutions and resources that will address these problems. Most importantly, it is the NGO's role to question the dominant social paradigm and to work toward creating a just, equitable, and sustainable society. Strengths generally associated with the NGO sector include:

- Strong grassroots links
- Field-based development expertise
- Ability to innovate and adapt
- Process-oriented approach to development
- Participatory methodologies and tools
- Long-term commitment and emphasis on sustainability
- Cost-effectiveness

Each NGO is an entity unto itself, with its own mission, style of leadership, and unique operating culture, and each adopts strategies based on those factors. Every community-based NGO would like nothing better than to accomplish its mission and thus work itself out of a job.

Successful NGOs know that they have three overarching ethical responsibilities to the communities they represent. They must maintain credibility by always speaking the truth; they must be fully accountable for their actions; and they must build the infrastructure to maintain the organization for the life of the issues they choose to address. Without fulfilling all three of these responsibilities, NGOs are subject to criticism and failure.

Ideally, given the needed resources, the NGO will take on various roles including that of educator, government and industry watchdog, community organizer, political activist, litigator, researcher, or even investor in a beneficial project—all roles that Amigos Bravos has played in efforts to clean up the Red River. More often than not, successful campaigns to address complex, chronic pollution issues—such as ones presented by the Molycorp mine—require multiple skills and knowledge that no single community-based organization can possess. In those instances, a number of NGOs with specialized skills will come together in a concerted campaign.

In New Mexico, where communities are spread over vast distances, rural populations are small, and financial resources are scarce, local NGOs have had to take on the critical role of protecting the communities' interests by focusing on one specific part of an issue, and letting other NGOs—often within the same community—take on other aspects of the issue. For instance, in Questa the Río Colorado Reclamation Committee has taken on oversight of the Superfund process, Concerned Citizens of Questa continues its forty-year effort to hold Molycorp accountable for contamination, Artesanos de Questa is creating sustainable economic development alternatives to mining, the Questa Environment and Health Coalition is helping to determine impacts of tailings and water contamination on residents' health, and Amigos Bravos is focused on restoring the Red River.

WHY ARE THERE SO MANY NGOS IN NEW MEXICO?

The answer lies in the fact that despite progressive advances in the protection of health, the environment, and social welfare by the passage of many federal and state laws, the government has produced regulations that often have been written with industry at the table. These regulations are complex and hard to enforce and often depend on the good will of industry. Some of the regulations take several years to implement. The ground water discharge permits required under the New Mexico Water Quality Act of the early 1970s are one good example. The regulations requiring discharge permits were adopted in 1978, yet Molycorp did not receive a ground water discharge permit for its mining operations until 1999.

Many regulations depend on voluntary compliance by industry. An example of this is the way permits are written and non-point source pollution is controlled. In New Mexico it is common practice for industry to suggest permit language that is then reviewed by the agency. Because of this, it falls on the NGO's shoulders to see that regulations ultimately include conditions that address community concerns. Molycorp's mining permit, in which Amigos Bravos played a key role, contains no fewer than sixty-four conditions. Similarly, New Mexico depends on voluntary compliance, through the development of best management practices (BMPs), to control acid rock drainage and other non-point sources of water contamination. Because BMPs can fall short of controlling pollution, it may fall to the community-based NGO to apply political pressure to ensure that the issues are addressed.

At the same time, the mining industry has been known to challenge the government's authority to enforce certain regulations. A prime example of this attitude is Molycorp's 1989 proposal for the Guadalupe Mountain tailings facility, in which the mine challenged the Bureau of Land Management's authority to require the mine to look at alternatives during the NEPA (National Environmental Protection Act) analysis. More recent examples include Molycorp's 1999 and Phelps Dodge's 2003 challenges before the New Mexico Water Quality Control Commission regarding how the state should regulate ground water at mining operations.

A further complicating factor for communities is the lack of consistency and continuity in government funding and policy implementation. As administrations change, so do policy and funding priorities, with the result that some initiatives are discontinued or left with greatly reduced funding. The present funding crisis for clean-up at Superfund sites is a prime example. All of these factors have fostered an increasing need for community stewardship of its resources—and hence a growth in the number of NGOs. The fact that there are so many NGOs working on mining issues points to a basic societal problem: Mines continue to produce waste, and, despite excellent laws and regulations, pollution continues to threaten public health and the environment.

To reverse this problem, NGOs must find ways of empowering regulators to claim the authority to hold the mining industry accountable to the communities where these natural resources are located. NGOs can bring a broad array of resources to the table that would otherwise not be available to regulators and industry. Besides monitoring data, scientific analysis, on-the-ground restoration efforts, and problem-solving expertise, NGOs often engage people from a broad spectrum of society who, though impacted by mining practices, would normally not be part of a mine-related, decision-making process—including people involved in the arts, human health, financing, etc. In order for NGOs to be effective voices for the concerns of their communities, and for society to benefit from an NGO's expertise, the NGO should be empowered to sit at the decision-making table whenever industry is present—especially when addressing environmental issues.

A CASE STUDY: AMIGOS BRAVOS

Incorporated in 1988, Amigos Bravos, Inc. is a statewide environmental NGO guided by social justice principles, with offices in Taos and Albuquerque, a staff of seven, and a membership of 1,500 individuals, families, and businesses. Because rivers are the lifeblood of New Mexico's communities, human and natural, Amigos Bravos works to protect both the ecological and cultural richness engendered by rivers. Amigos Bravos accomplishes its mission through direct advocacy and by empowering individuals and communities to protect local water resources.

It is the mission of Amigos Bravos to:

- Return New Mexico's rivers and the Río Grande watershed to drinkable quality wherever possible, and to contact quality everywhere else

- Ensure that natural flows are maintained and, where those flows have been disrupted by human intervention, to advocate that they be regulated to protect and reclaim the river ecosystem by approximating natural flows

- Preserve and restore the native riparian and riverine biodiversity

- Support the environmentally sound, sustainable, traditional ways of life of indigenous cultures

- Ensure that environmental justice and social justice go hand in hand

Amigos Bravos uses all advocacy tools including educating the public through the media and other venues; offering opportunities for volunteer action; working with policymakers to enforce environmental laws and to adopt progressive policies; participating in complex administrative and regulatory processes including standard-setting, rulemaking, and permitting; and undertaking protection and restoration efforts.

At its inception, Amigos Bravos became active on mining issues in response to the pollution of the Red River from many tailings pipeline spills, and the consequences of operations at the Molycorp mine. Subsequently, when Molycorp proposed to build a tailings facility on the saddle of Guadalupe Mountain, adjacent to the Wild Rivers Recreation Area, Amigos Bravos joined many other NGOs in opposing the construction of the tailings facility. Since then Amigos Bravos has been involved in many mining-related issues and campaigns dealing with both acute and chronic situations.

THE ROLE OF THE NGO IN ACUTE VS. CHRONIC SITUATIONS

The proposal to develop a new tailings facility and the ongoing contamination of the Red River represent two very different situations requiring different NGO roles, strategies, and attributes. Whereas a proposed new tailings facility in an unsuitable location presents an acute problem requiring a fast and immediate response with maximum public input, the contamination of the Red River represents a chronic problem requiring a long-term clean-up strategy that may go on beyond the lifetime of the individuals involved at the beginning.

In response to the acute problem, NGOs abide by the belief that decision makers have to respond to public outcry—although this may not be what actually happens, because of legal, political, and/or security and safety reasons. Existing NGOs, within and outside of the community, mobilize their membership and dedicate resources in order to achieve a quick resolution. Newly formed NGOs may have to respond to the various elements and/or requirements of the crisis. Once the issue has been resolved, the newly formed NGOs often go dormant until a similar issue arises again. Local NGOs will go back to their ongoing work, and regional and national NGOs will go on to the next acute battleground challenge.

Dealing with chronic pollution situations requires a much broader combination of roles and challenges for the NGO. Chronic situations usually involve large quantities of toxic substances, and a complex set of physical and legal circumstances. All too often politicians and the public grow tired of dealing with long-term problems and, because of the complexity of the issues, feel powerless to do anything about them. In those situations the NGO has to develop and maintain community support; credibility with the media, the public, and the regulators; staying power for the long haul; technical expertise to provide solutions; a place at the negotiation table; and a role in the implementation and enforcement of cleanup plans. For Amigos Bravos, each of these attributes has represented a new stage in the development of the organization—and to a great extent has required that the organization take on new roles and new challenges.

I will conclude with some personal observations and recommendations regarding the role that NGOs should be able to exercise in the decision-making process.

The NGO should be recognized as a legitimate voice for concerns, and a valuable asset to the decision-making process. Too often, I have come away from meetings with decision makers feeling ignored and discounted. I have heard the retort that NGOs do not represent communities—that only elected officials hold that privilege. Decision makers should recognize that NGOs speak from convictions that transcend political agendas. NGOs raise concerns of individuals, species, and natural systems that often do not have a voice at the table. In many instances NGOs are the canaries in the mineshaft.

Moreover, some NGOs are perceived as troublemakers or obstructionists. The responsible NGO is interested in finding sustainable solutions to acute and chronic problems. NGOs will often propose creative solutions well in advance of industry and government —who are constrained by their own culture and perceptions. It is the decision-makers' responsibility to explore NGO alternatives.

NGOs need the necessary resources to fully participate in the decision-making process. NGOs provide a tremendous range of expertise, experience, and information that would not otherwise be available to the decision maker. It often falls upon NGOs to demand accountability, bring to the discussion sensitive community information, propose alternatives, and/or counter industry's experts. NGOs add value to the regulatory process that should be recognized, and the added costs accepted as part of doing business. It is in the long-term interest of industry and government to ensure that NGOs in the affected communities are able to participate fully in the decision-making process. Society as a whole—regulators and industry included—will harvest great benefits from NGO participation.

NGOs should be invited into the decision-making process. This requires that regulators exert their authority to provide adequate notification to the communities affected by mining. Notices in newspapers and radio are not enough, especially in low-income and minority communities where personal contact is the primary way of engaging residents. In those situations, regulators and industry should contract with community organizers to engage key community representation and expertise. In the interest of creating a healthy and sustainable future, it behooves decision makers to create opportunities for NGOs to be at the decision-making table.

A Partial List of Public Interest NGOs Working on Mining Issues in New Mexico

Amigos Bravos
Big Mountain
Black Mesa Trust
Black Mesa Water Coalition
Carson Forest Watch
Center for Biological Diversity
Center for Science in Public Participation
Citizens Coal Council
Coalition for Clean Affordable Energy
Concerned Citizens of Questa
Diné Care
Diné Mining Action Center
EARTHWORKS
Eastern Navajo Uranium Workers
ENDAUM—Eastern Navajo Diné Against Uranium Mining
Friends of Santa Fe County
Forest Guardians
Gente del Río Pecos
Gila Resources Information Center
Laguna-Acoma Coalition for a Safe Environment
Mining Impacts Communications Alliance
Moquino Domestic Water Users Association
New Mexico Citizens for Clean Air & Water
New Mexico Conservation Voters
New Mexico Environmental Law Center
New Mexico Mining Act Network
New Mexico Public Interest Research Group
Questa Environment and Health Coalition
Río Colorado Reclamation Committee
San Juan Citizens Alliance
Sierra Club
Southwest Research and Information Center
Trout Unlimited
Vecinos del Río
Water Information Network
Western Environmental Law Center
Western Mining Activists Network
Westerners for Responsible Mining
The Wilderness Society
Zuni Salt Lake Coalition

The Future of Mining in New Mexico

The mining industry has played and continues to play an important role in the economic prosperity of New Mexico. In 2003 there were 225 active mining operations in the state, employing more than 5,000 people, with an annual payroll of $229 million. That same year the industry generated over $27 million in revenues to the state of New Mexico. But what of the future of mining in New Mexico? We as a nation continue to rely upon a viable mining industry to support a quality of life that we take for granted. But that future, here and elsewhere, will rely on achieving a balance between our needs and desires, the changing economy, and our growing concern over environmental and social issues that face all of us (many of which are not unique to the mining industry). We cannot predict what role mining will play in the future, but we can state very clearly that it will depend upon our ability to face a number of very specific challenges down the line.

THE REMAINING RESOURCE POTENTIAL IN NEW MEXICO

The single most important factor that will determine the future of mining in New Mexico is the presence of economic mineral deposits in the state. New Mexico is at the eastern edge of one of the world's great metal-bearing provinces. New Mexico (and the neighboring states of Utah and Arizona) are host to significant, even world-class copper deposits. The potential for discovery of additional metals in New Mexico is good.

Yet, despite the recent increase in commodity prices, exploration and production of mineral deposits in New Mexico are at an all time low. Most known mineral deposits in New Mexico are currently being mined or are not economic at present, because they are either too small, too low grade, too deep, or are in areas where it would be difficult to mine. The Jones Hill project north of Pecos in the Santa Fe National Forest is one of several small, medium-grade mineral deposits containing gold, silver, lead, zinc, and copper. But a permitted mine there would require the costly construction of haul roads through forested mountains that are popular recreational areas, and it could have environmental impacts on the Pecos River. Similar deposits in the area are even less likely candidates for mining because they are in rugged, undeveloped canyons. Some of the known mineral deposits that might be considered favorable exploration targets are found in or near Wilderness or Wilderness Study Areas (such as deposits in the Nogal area in Lincoln County), on restricted lands, or on land withdrawn from mineral entry (such as deposits in the Elizabethtown–Baldy district in Colfax County).

Economically it will be very difficult to mine at this time the lead and zinc that exist in New Mexico. Vast reserves of high-grade, near-surface lead and zinc ore exist in other states (Cominco's Red Dog deposit in Alaska and Doe Run Company's deposits in the Viburnum Trend in Missouri are two examples). New Mexico's lead and zinc deposits are low grade, small tonnage, deep, and costly to mine and process. Existing lead and zinc smelters are far from New Mexico. Anyone wanting to mine lead and zinc in New Mexico would have a difficult time finding a buyer for the concentrate.

The closure of the Hurley (Grant County), Hidalgo (Playas, Hidalgo County), and ASARCO (El Paso) smelters has prevented some small metal mines from operating. Several small mines in the Lordsburg, Steeple Rock, Mogollon, and Hillsboro districts have operated in the past twenty years only by producing silica-fluxing ores containing gold, silver, and copper. These mines did not need mills but depended upon the smelters purchasing their crushed, unprocessed ores. These small metal mines are typically too small and low grade to finance the construction of a mill, and without a ready market they cannot exist strictly as metal mines.

Potential exists for finding economic gold deposits in several areas in New Mexico. However, some of the highest-grade deposits are at Ortiz in Santa Fe County, and a number of environmental factors have prevented any mining there in recent years. Remaining gold deposits in New Mexico would be classified as exploration targets rather than ore deposits.

New Mexico is second only to Wyoming in uranium reserves in this country. It will be difficult to mine uranium by conventional methods in New Mexico because in-state deposits currently have difficulty competing with production from Canada and Australia, and there are no processing facilities in New Mexico at this time. At the same time, there are several companies actively exploring for uranium in New Mexico today. The industry is exploring the possibility of mining existing reserves in place through the use of in situ leaching techniques. Whatever the outcome, it is very likely that there will be renewed interest in the state's uranium resources in the future.

New Mexico is third in the nation in terms of coal reserves, and industrial minerals are an increasingly significant commodity in the state. New Mexico is first in the nation in the production of potash, perlite, and zeolite. Both limestone and aggregate are available in abundance. Yet 20 percent of the cement we use in the U.S. is currently imported, and in New Mexico there remains only one active cement plant. There are concerns today regarding pressures to impose new regulations on the sand and gravel industry. The primary concern is over the lack of a statewide reclamation requirement for sand and gravel operations. Regardless of the outcome, regulation will likely affect the future of the sand and gravel industry in New Mexico.

Virtually all mineral deposits are mined only when a mining company believes that the operation will make a profit under prevailing market conditions. One exception to this is strategic minerals, which are those minerals deemed critical to military, industrial, or civilian needs during a national emergency. Under such conditions, some minerals could become economically viable overnight. The sources upon which the U.S. relies for many strategic minerals are often found in politically unstable countries. Over 90 percent of the world's platinum, manganese, and chromium deposits are found in the former Soviet Union and South Africa. During World War II, deposits such as those at the Harding mine were developed for this purpose.

Although strategic minerals are not mined today in New Mexico, there is the potential for these deposits to be tapped in the future. For example, manganese production from New Mexico until 1981 amounted to 1,750,096 long tons, and remaining reserves in small, low-grade deposits are abundant in some districts including Luis Lopez, Deming, and Boston Hill. The Taylor Creek district is one of the few areas where tin (used in bronze and brass) is found in the United States. Other small, low-grade deposits of strategic mineral materials including beryllium, iron, nickel, cobalt, titanium, tungsten, barite, and tantalum are found throughout the state. They are available for production if and when they are needed, should foreign sources become unreliable.

THE INFLUENCE OF THE REGULATORY FRAMEWORK

The citizens of New Mexico play an important part in deciding how or if mining is conducted here in the future. They do this by supporting the passage and enforcement of laws and regulations designed to control where, how, when, and ultimately if mines go into operation. The tide of public opinion can (and does) change, and lawmakers follow suit.

Laws are passed to address or prevent specific negative impacts from mining such as water or air contamination, but in the end they have more far-reaching consequences. If the costs a mining company must incur to comply with regulations make the operation unprofitable, then the regulations have prevented the mine from going into operation. The negative environmental and social impacts are avoided, but the minerals, jobs, revenues, and other economic benefits that would have accompanied them are lost. The current trend in public policy and regulation is toward allowing mining only if it can be done in an environmentally responsible manner that limits factors such as visual impact, truck traffic, noise, dust, and other health, safety, and environmental impacts. Perhaps the three most significant elements that have been incorporated in recent laws are increased public involvement, requirements to reclaim mines to other beneficial land uses after closure, and financial assurance requirements to ensure permit requirements are met.

Mining companies weigh the regulatory burden that applies to potential mines in different states and countries along with many other factors when deciding which options to pursue. Where does New Mexico stand compared with other states? In 2004 the Fraser Institute, an independent Canadian economic and social research organization, published a report that summarizes the opinions of many mining company executives and compares certain states and countries in terms of several factors related to mining today. This report places New Mexico fourth out of fourteen western states in terms of which state's environmental regulations are favorable to exploration investment, ranking ahead of states like Arizona and Alaska. Only Nevada, Utah, and Wyoming were considered more favorable. New Mexico ranked somewhere in the middle for most regulatory and public policy-related factors. When other countries are included in the mix, our state still ranked in the middle. Nevada consistently ranked as the most desirable state in the U.S., whereas Wisconsin and California consistently ranked as the least desirable. The methodologies and accuracy of the Fraser Institute report are not without controversy, but it is one way of taking a broad look at the industry worldwide. Although individual companies' opinions vary considerably, in general the current regulatory framework does not seem to strongly dissuade companies from coming to New Mexico. Other factors such as mineral resource potential play a larger role.

Ultimately the consumer pays for the added costs associated with complying with regulations, because these costs are included as a part of doing business. If the citizens of New Mexico wish to encourage mining, steps can be taken within the regulatory process to facilitate such development. For example, it can take many years to obtain all necessary federal and state permits to open a major mine today. This is one of the problems most commonly cited by mining companies operating in New Mexico. If this process can be streamlined and multiple agency requirements coordinated, this time period could be substantially reduced. Resource protection and land-management obligations could be simplified and consolidated into fewer programs without weakening the provisions. Through legislation and regulation, tax burdens could be cut, development incentives initiated, and infrastructure assistance provided.

Conversely, if our populace decides mining is not in the best interest of New Mexico, it can deter or effectively eliminate active mining through passage of more stringent laws, or by placing a moratorium on mining that will send mining companies to more receptive locations. Perhaps more likely than either of these two extremes is a scenario where laws and regulations are refined to reduce environmental degradation and provide citizen protections, while not preventing mining altogether. Other states and countries are likely to do the same.

ECONOMICS

Geology determines where a deposit occurs and what methods must be used for extraction. The transportation network connecting the buyer with the seller finally determines the delivered price. Beyond these basics, mining is a very competitive, worldwide, cyclic business characterized by low return on equity and long lead times to first production.

The current high mineral prices are based primarily on the rapid increase in demand, as a result of the high annual economic growth that has occurred in China, India, and other countries over the past several years, as well as a decline in the value of the U.S. dollar. China has traditionally been an exporter of many minerals, offering high grade and low price, which drove many operations out of business worldwide. China now uses much of its production internally, and is importing as well. This decrease in mineral supply and increase in demand has caused prices to increase worldwide. Increased prices bring former operations back into production, which ultimately increases supply and lowers price—perpetuating the classic boom-bust cycle of the mineral business.

In the U.S., most deposits are developed over a period of ten to twelve years, due mainly to the permitting process. This period is far longer than the mineral supply/demand cycle. Predicting mineral price and demand that far in advance is difficult, so developing a mineral deposit is financially risky. Coming online in a down cycle can seriously hamper or doom an operation. Risk is inherent in the exploration process, because only a few of the deposits evaluated are developed, and most exploration funds are spent on undeveloped prospects.

About $1.2 billion was spent worldwide in 2004 for mineral exploration. Overall mining capitalization (investments) is about $390 billion, largely controlled by international companies that are based mostly in Europe. The return on equity for U.S. mining was 10 percent in 2004. This return is up from many years of 6–7 percent, which is close to the return on equity that can be realized without such risk. This low return is similar to railroads (8.54 percent in 2004, a banner year), which have been in financial difficulties for decades. By comparison, the return for U.S. industry as a whole in 2004 was about 12 percent.

Given the narrow margins under which these companies operate, the cost of complying with financial assurance and bonding requirements severely affects the ability of these companies to see a return on their investment. Financial assurance requirements have risen dramatically in recent years and have a significant effect on the willingness of companies to do business here. New Mexico has imposed financial assurance requirements on several mining operations in the state that are among the highest in the world.

Regionally, New Mexico is in a similar situation for

mineral exploitation as surrounding states. Although New Mexico has somewhat higher taxes, most western states are viewed as relatively unattractive exploration targets by mining executives, as expressed in the 2004 Fraser Institute study. The transportation network in New Mexico has pluses and minuses. On the plus side, trucking backhauls are abundant and inexpensive because New Mexico is not a manufacturing state, so more truckloads arrive than depart. The rail network is adequate, but a spur line into the northwestern coal fields would benefit New Mexico greatly. On the negative side, many areas with mineral potential are served only by gravel roads, which must be upgraded, adding significantly to the cost of mine development. Access to shipping by water is primarily through the ports of Los Angeles and Houston but requires truck or rail haul to those cities, which partly eliminates the savings associated with water transport. Overall, New Mexico mines serving local markets are partially protected from outside competition by the transport network, so the relative remoteness of the state works both ways. But mineral sales to distant markets are easily lost to competitors who have easier access to lower-cost transportation.

THE CHALLENGES OF THE FUTURE

What would it take to breathe new life into the mining industry in New Mexico, or even to maintain the status quo? Clearly a lot will depend upon increasing demand, streamlining the regulatory framework, and a favorable economic environment. Yet it is clear that the mining industry must be accountable to growing concerns over environmental issues. It is difficult to predict how all of these challenges will resolve themselves, but ultimately the future will depend upon the following:

- Water. There is not a single industry in New Mexico whose future does not depend upon both access to water and the ability to guarantee protection of the quality of the state's water resources.

- Environmental concerns. There must, indeed, be a way to do it right, and the public increasingly demands this.

- A willingness on the part of the population to say, "Yes, in my back yard, if our economic well-being and way of life depend upon it. But with a few caveats."

- The mining companies must be good neighbors. And this has to do both with perception and with reality.

- The issue of sustainable development. The mining industry must find a way to fit into this developing model, which has more to do with *how* the industry does its business than *where*, and for *how long*.

- The nature of the regulatory framework. Although laws are sometimes designed to be obstructive, the process itself should not be.

Many of these challenges are not unique to this industry; they reflect the realities of our time. Facing these challenges successfully will depend upon a willingness to communicate and collaborate, a reliance on the best that science can offer, and an inherent concern with both the environment and our quality of life.

—*The Editors*

List of Contributors

Barker, James M.
Senior Industrial Minerals Geologist/Associate Director
New Mexico Bureau of Geology and Mineral Resources
New Mexico Institute of Mining and Technology
801 Leroy Place
Socorro, New Mexico 87801
(505) 835-5114 Fax: (505) 835-6333
jbark@nmt.edu
James Barker is a senior industrial minerals geologist and associate director at New Mexico Bureau of Geology and Mineral Resources and adjunct faculty member, Department of Earth and Environmental Science and Department of Mineral Engineering. He studies the important roles that marketing and transportation play in successful industrial mineral ventures. His work at the bureau has included development of the perlite lab, which is the main perlite testing facility in North America. He has studied in detail the state's deposits of perlite, zeolite, potash, aggregate, pumice, humate, barite, cement, dimension stone, and others. Before joining the bureau, he taught geology and worked as a geologist for several mining companies. Barker has written and edited well over two hundred articles, papers, reports, books, posters and abstracts on industrial minerals. James is an honorary member in the New Mexico Geological Society and a distinguished member of Society for Mining, Metallurgy, and Exploration. Barker received a B.S. in Geology from the University of California, Los Angeles, and an M.A. in Geology from the University of California, Santa Barbara.

Barnes, Julia H.
Office of Public Facilitation
200 West De Vargas, Suite 2
Santa Fe, NM 87501
(505) 982-3993
jhb1@nm.net
Julia Barnes is currently working for the state of New Mexico as a staff member of the Office of Public Facilitation. This state government office works with state agencies to promote the use of collaborative processes, particularly on larger public policy issues. The office is administratively attached to the Environment Department. Since its inception in January 2002, the Office of Public Facilitation has worked with all three branches of government, in many departments and divisions of state government. Barnes trains state personnel and court volunteers in mediation and facilitation techniques in addition to providing process expertise to agencies. Barnes has been a practicing attorney since 1989, a mediator since 1990, and a facilitator since 2000. Barnes received an undergraduate degree from Stanford University and a law degree from the University of Colorado at Boulder.

Bauer, Paul W.
Senior Geologist/Associate Director
New Mexico Bureau of Geology and Mineral Resources
New Mexico Institute of Mining and Technology
801 Leroy Place
Socorro, NM 87801
(505) 835-5106 Fax: (505) 835-6333
bauer@nmt.edu
Paul Bauer is a senior geologist and associate director at the New Mexico Bureau of Geology and Mineral Resources and adjunct faculty member, Department of Earth and Environmental Science. He has spent twenty-five years investigating the sublime geology and landscapes of New Mexico. His current research is focused on using field studies and structural geology to describe the geologic evolution of the mountains and basins of north-central New Mexico. He has co-authored a series of geologic quadrangle maps of southern Taos County and has published a wide variety of books and articles on the geology of the region. He is Program Coordinator for the Decision-Makers Field Conferences, a five-year-old program that attempts to bridge the gap between science and policy in New Mexico. Paul received a B.S. in geology from the University of Massachusetts, an M.S. in geology from the University of New Mexico, and a Ph.D. in geology from New Mexico Institute of Mining and Technology.

Bland, Douglas M.
Economic Geologist
New Mexico Bureau of Geology and Mineral Resources
New Mexico Institute of Mining and Technology
801 Leroy Place
Socorro, NM 87801
(505) 466-6696 Fax: (505) 466-3574
douglasbland@msn.com
Doug Bland has worked for the New Mexico Bureau of Geology and Mineral Resources since 2004, primarily as technical coordinator and trip leader for the Decision-Maker's field conference. He served as director of the Mining and Minerals Division of the New Mexico Energy, Minerals and Natural Resources Department from 1998 through 2002, where he was responsible for overseeing environmental protection and permitting of mine sites under the New Mexico Mining Act and Coal Surface Mining Act. He also held various technical and managerial positions in Mining and Minerals Division between 1989 and 1998. Hhis experience includes twelve years in the mining and petroleum industries. He holds B.S. and M.S. degrees in geology from Virginia Tech and the University of Wyoming.

Brancard, William
Director, Mining and Minerals Division
Energy, Minerals and Natural Resources Department
1220 South St. Francis Drive
Santa Fe, NM 87505
(505) 476-3405
bbrancard@state.nm.us
Bill Brancard has worked extensively on New Mexico mining issues for more than a decade as a government official and as a public and private attorney. Brancard is the governor's representative to the Interstate Mining Compact Commission and the Western Interstate Energy Board Reclamation Committee. He was the first general counsel to the New Mexico Mining Commission when the commission was created in 1993 and later served as vice chair of the commission. Brancard was an assistant attorney general from 1991 to 1997, and from 1997 to 1999 he was an Assistant Land Commissioner for Policy and Legislation at the State Land Office. Before attending law school, he worked as an archaeologist in New Mexico and Arizona and did graduate studies in archaeology at the University of New Mexico. He has been listed in *Who's Who in American Law* and has published legal articles in the Harvard Journal on Legislation and the New Mexico Law Review. Brancard has a B.A. in anthropology and history from Hamilton College, a J.D. from Harvard Law School and also attended the London School of Economics.

Coleman, Michael W.
Watershed Protection Team Leader
Surface Water Quality Bureau
New Mexico Environment Department
P.O. Box 26110
Santa Fe, NM 87502
505-827-0505
michael_coleman@nmenv.state.nm.us
Michael Coleman has been a geoscientist in the Surface Wager Quality Bureau, New Mexico Environmental Department since 1994. He has worked as a geologist, surface mine reclamation specialist, and environmental specialist. Before joining the Surface Water Quality Bureau, he worked tewnty-two years in the mining industry. He has extensive local and regional geologic and structural mapping, reconnaissance, and resource evaluation experience, and has participated in numerous energy, precious metals, and industrial minerals exploration and development projects, including all levels of reclamation work. In his current position as Watershed Protection team leader, his assignments include Clean Water

Act Section 319 grant project management, mine permitting and reclamation, watershed and stream restoration efforts, and he is presently actively involved in innovative erosion control and stream geomorphology projects in the challenging Rio Puerco Watershed. He received his geology degre from the University of New Mexico, and has done graduate work at UNM and the University of Nevada's Mackey School of Mines, in Reno.

D'Esposito, Stephen
President/CEO
EARTHWORKS
1612 K St., NW, Suite 808
Washington, D.C., USA 20006
(202) 887-1872 ext. 203
sdesposito@earthworksaction.org
Stephen D'Esposito became president/CEO of EARTHWORKS in January 1998. Before joining EARTHWORKS Steve served as vice president for policy and, from 1986 through 1992, was instrumental in building Greenpeace USA into one of the largest environmental groups in the U.S. From 1993 through early 1996, Steve served as deputy director and then executive director of Greenpeace International in Amsterdam, the Netherlands. During his tenure at Greenpeace International he launched the Brent Spar campaign to block Shell from dumping oil facilities in the North Sea. Before his work with Greenpeace, Steve was field director for the New York Public Interest Research Group. Steve received a bachelor's degree in political science from Tulane University in New Orleans.

Eveleth, Robert W.
Senior Mining Engineer
New Mexico Bureau of Geology and Mineral Resources
New Mexico Institute of Mining and Technology
801 Leroy Place
Socorro, NM 87801
(505) 835-5325 Fax: (505) 835-6333
beveleth@gis.nmt.edu
Robert W. Eveleth is associate curator of the bureau's mineral museum; manager of the Bureau's archival collections of mining and mineral-related data, photographs, manuscripts, and maps; and, since 1986, chairman or co-chairman of the annual New Mexico Mineral Symposium at Socorro. He also works closely with the New Mexico Abandoned Mine Lands Investigative Team. His experience includes substantial underground mining and open pit with various base and precious metal mines in the southwestern U.S. He has conducted field studies of many New Mexico mines and districts, provided technical advice to small mine operators, prospectors, and exploration geologists; and been principal investigator on several data base projects for the U.S. Bureau of Mines and U.S. Geological Survey. He co-authored the U.S. Bureau of Mines New Mexico chapter in the Mineral Yearbook for many years. He received his degree in mining engineering from the New Mexico Institute of Mining and Technology.

Evetts, Robert M.
P.O. Box 1124
Tucumcari, NM 88401
Robert Evetts is a civil engineer and has been the Abandoned Mine Land program manager in the New Mexico Mining and Minerals Division since April 1990. Evetts was the deputy secretary of the Energy, Minerals and Natural Resources Department in 1989. He is a licensed contractor in New Mexico and proprietor of R. M. Evetts Engineering and Construction in Tucumcari, New Mexico, founded in 1972. He has been listed in the Who's Who In American Pipeline Contractors and is a member of the New Mexico State University Academy of Civil Engineering. He was selected as a New Mexico State University Centennial Engineering Alumni in 2000. He is a licensed professional engineer in New Mexico, Colorado, and Texas. Evetts received his B.S. degree in Civil Engineering from New Mexico State University. He retired in 2004.

Foreback, Terence G.
Senior Environmental Engineer
Mining and Minerals Division
Energy, Minerals and Natural Resources Department
1220 S. St. Francis Dr.
Santa Fe, NM 87505
(505) 476-3432
Terence.Foreback@state.nm.us
Terence Foreback has current expertise in engineering activities related to the Mining and Minerals Division, Mining Act Reclamation Program including permitting, reclamation, slope stability, and financial assurance. Before joining the Mining and Minerals Division, he held various positions in the mining industry including general reclamation supervisor, general blasting supervisor, mine supervisor, and resident mine engineer. Terence received his B.S. degree in mining engineering from the Pennsylvania State University and is a registered professional engineer in Colorado and New Mexico.

Freeman, Patrick S.
President, St. Cloud Mining Company
1243 Marie Street
P.O. Box 1670
Truth or Consequences, NM 87901
(505) 743-5215 Fax: (505) 743-3333
stcloud@riolink.com
Pat Freeman is president of St. Cloud Mining Company where he has been employed since 1980. St. Cloud has explored, operated, and reclaimed several surface and underground base and precious metal mines in New Mexico and continues to produce natural zeolites and construction aggregates. The company also provides construction and mine reclamation services for the state of New Mexico and private concerns. He began his career as an underground mine geologist at the San Pedro Mine in southern Santa Fe County in 1968, worked domestically and internationally for Texasgulf, Inc. in mine evaluation and development. Freeman served as an industry representative on the New Mexico Mining Commission from 1996–2002. He is a member of the American Institute of Mining Engineers and a certified professional geologist. Freeman received his B.A. in geology from Monmouth College.

Garcia, Karen
Mining and Minerals Division
Energy, Minerals and Natural Resources Department
1220 South St. Francis Drive
Santa Fe, NM 87505
(505) 476-3435 Fax: (505) 476-3402
kwgarcia@state.nm.us
Karen Garcia is a bureau chief with the Mining and Minerals Division, overseeing both the Mining Act Reclamation Program and the Coal Mine Reclamation Program. She started as a reclamation specialist with the Mining Act Program in 1999. Karen has a degree in wildlife biology and management from New Mexico State University.

Hall, E. L. (Ned)
Phelps Dodge, New Mexico Operations
210 Cortez Ave.
P.O. Box 7
Hurley, NM 88043
(505) 537-4237 Fax: (505) 537-8012
ehall@phelpsdodge.com
Ned Hall has been with Phelps Dodge Mining Company's New Mexico Operations since 1994. He is currently the manager of the Environment, Land and Water Department and is responsible for reclamation and closure activities for Phelps Dodge's Chino, Tyrone and Cobre Mines in southwest New Mexico. Ned received his B.S. degree in engineering science from New Mexico Institute of Mining and Technology and his M.S. in environmental engineering from New Mexico State University

Harben, Peter
President, Peter W. Harben Inc.
Industrial Minerals Consultants
5251 Eagle Pass Road

Las Cruces, NM 88011
(505) 521 3301 Fax: (505) 522 5389
peter@peterharben.com
Peter Harben has participated in more than two hundred assignments involving the market evaluation of the full spectrum of industrial minerals, certain inorganic chemicals, and selected metals. The work has included technical visits to over forty countries on behalf of clients ranging from major multi-nationals and the World Bank to individual landowners. Projects have included worldwide market evaluations, assessment of acquisition opportunities, the development of marketing plans, and participation in due diligence work. The firm works closely with professional associates whose expertise includes transportation, health and safety regulations, mineral processing, financial analysis, and glass technology. A former American editor of Industrial Minerals, Peter Harben has written and spoken extensively on the subject of industrial minerals and their markets. Texts include Industrial Minerals: A Global Geology (three editions), The Industrial Minerals HandyBook—A Guide to Markets, Specifications and Prices (four editions), Manganese—Uses and Markets, Bauxite and Alumina—Uses and Markets, Sampling Industrial Minerals and Ores, and the titles Kaolin, Salt, Beryllium, and Industrial Diamonds in a series of Financial Times Executive Commodity Reports.

Hogge, David
Program Manager, Monitoring and Assessment Section
Surface Water Quality Bureau
New Mexico Environment Department
1190 St. Francis Drive
P.O. Box 26110
Santa Fe, NM 87532
(505) 827-2981 Fax: (505) 827-0160
david_hogge@nmenv.state.nm.us
David Hogge monitors and assesses surface water quality in New Mexico's rivers, lakes, and streams to determine if state surface water quality standards are met and to ensure that designated uses are supported. This effort includes development of total maximum daily loads for impaired waters and maintenance of an extensive database and Geographic Information System on the quality of the state's surface waters. David has worked thirteen years in the environmental field including work with Lockheed Environmental Systems and Technologies Company, Las Vegas, Nevada, contract laboratory program from 1990 to 1994. He holds a B.S. degree in community and public health and an M.S. degree in health administration from New Mexico State University.

Johnson, Peggy
Senior Hydrogeologist
New Mexico Bureau of Geology and Mineral Resources
New Mexico Institute of Mining and Technology
801 Leroy Place
Socorro, NM 87801
(505) 835-5819 Fax: (505) 835-6333
Peggy Johnson is a senior hydrogeologist with the New Mexico Bureau of Geology and Mineral Resources. She is currently working with the Office of the State Engineer on the hydrogeology of the Española Basin near Santa Fe. Results of one of her recent studies on the hydrogeology and water resources of the Placitas area of north central New Mexico provide the scientific framework for the area's regional water planning effort. She has eighteen years of consultant and research experience in ground water hydrology and related fields. Her diverse background includes practical research in basin hydrogeology, karst hydrology, mountain-front recharge, surface water and ground water resource assessments, isotope hydrology, and water resource management and policy. Johnson has considerable previous experience in private consulting and conducts hydrogeologic and water supply investigations for the New Mexico Office of the State Engineer, the Interstate Stream Commission, and various counties and municipalities throughout the state. She received her B.S. in geology from Boise State University (Idaho), and her M.S. in hydrology from New Mexico Institute of Mining and Technology.

King, Gary
P.O. Box 1209
Carlsbad, NM 88221
and
P.O. Box 40
Moriarty, NM 87035
(505) 832-4461 Cell: (505) 238-9464
lglgking@nm.enet
Gary King is in the private practice of law, specializing in environmental issues, water law, and land use planning. He previously served as the director of the Office of Worker and Community Transition at the U.S. Department of Energy (DOE), and as policy advisor to the DOE assistant secretary for environmental management. Before joining the DOE, he served as general counsel and senior environmental scientist for Advanced Sciences Inc. He served six terms in the New Mexico House of Representatives where he was the primary sponsor of legislation involving air and water quality, hazardous waste transportation and disposal, mine reclamation, and other environmental issues. In 2004 he was the Democratic nominee for Congress in the Second District of New Mexico. King is currently the chair of the New Mexico Mining Commission. He holds a B.S. degree in chemistry from New Mexico State University and a Ph.D. in organic chemistry from Colorado University. Dr. King earned his law degree at the University of New Mexico.

Kuipers, James R.
Kuipers & Associates
P.O. Box 641
739 W. Broadway
Butte, MT 59703
(406) 782-3441 Fax: (406) 782-3351
jke@montana.com
Jim Kuipers is a mining environmental consultant and the principal and a consulting engineer with Kuipers and Associates based in Butte, Montana. He has worked on mining and environmental projects including engineering design, permitting, operations, reclamation and closure, and financial assurance for over twenty years. Since 1996 his primary work has been as a consultant providing expertise to public interest groups, state, federal, and tribal governments, and industry relative to mining environmental issues. Previously he worked as a miner, engineer, and manager in the hard rock mining industry. Jim is a registered professional engineer in Colorado and Montana. He received a B.S. from Montana College of Mineral Science and Technology in mineral process engineering.

Langer, William H.
U.S. Geological Survey
MS 964 P.O. Box 25046
Denver, CO 80225-0046
(303) 236-1249 Fax: (303) 236-1409
blanger@usgs.gov
Bill Langer has been a research geologist with the U.S. Geological Survey since 1971 and has been the U.S. Geological Survey resource geologist for aggregate since 1976. He is a member of the Society for Mining, Metallurgy, and Exploration, the American Society for Testing and Materials committees for Concrete Aggregate and Road and Paving Materials, and the International Association of Engineering Geologists Commission No. 17 on aggregates. Bill has conducted geologic mapping and field studies of aggregate resources throughout much of the United States. He has published over one hundred reports, maps, and articles relating to crushed stone and gravel resources including monthly columns about geology and aggregate resources in *Aggregates Manager* and *Quarry*.

McCullough, Warren
Bureau Chief, Environmental Management Bureau
Montana Department of Environmental Quality
1520 East Sixth Avenue
P.O. Box 200901
Helena, MT 59620-0901
(406) 444-6791

wmccullough@state.mt.us

Warren McCullough is currently chief of the Montana Department of Environmental Quality's Environmental Management Bureau, which oversees the Metal Mine Reclamation Act and the Major Facilities Siting Act. He started working for the state in 1995 as a reclamation specialist/exploration geologist. Before that he accumulated over twenty years of industry experience in mineral exploration, oil and gas exploration and production, and corporate management. He has a B.A. in anthropology from Yale University, an M.S. in economic geology from Bowling Green State University (Ohio), and has studied at the University of Edinburgh (Scotland), University of Utah, Oklahoma State, and Montana Tech.

McLemore, Virginia T. (Ginger)
Senior Economic Geologist/Minerals Outreach Liaison
New Mexico Bureau of Geology and Mineral Resources
New Mexico Institute of Mining and Technology
801 Leroy Place
Socorro, NM 87801
(505) 835-5521 Fax: (505) 835-6333
ginger@gis.nmt.edu

Ginger McLemore came to the bureau in 1980 as a minerals geologist specializing in uranium deposits. Ginger is currently the lead investigator for the bureau in a multi-million-dollar project at Molycorp's Questa mine, on the characterization of the Questa rock piles. This cooperative project includes bureau staff, New Mexico Tech faculty and students, the University of Utah, and several other universities. The goal is to better understand the weathering processes that affect mine rock piles to mitigate possible hazards. She has published over one hundred and fifty articles on the metallic mineral resources of New Mexico, including her best-selling map of the gold and silver deposits of New Mexico. Ginger is an energetic researcher in the field of alkaline magmatism, carbonatites, anorthosites, and A-type granites. In 2004 Ginger received an Appreciation Award from the New Mexico Mining Association for her participation at their sixty-fifth annual meeting and the Team of Excellence Award from the Utah Engineering Experiment Station for her contributions to the Molycorp project. She also was elected chairman of the steering committee for the Acid Drainage Technology Initiative group, which conducts research and disseminates technologies on acid drainage. She holds B.S. and M.S. degrees in geology from New Mexico Institute of Mining and Technology and received her Ph.D. in geoscience from the University of Texas at El Paso.

Menetrey, Mary Ann C.
Program Manager, Mining Environmental Compliance Section
New Mexico Environment Department
P.O. Box 26110
Santa Fe, NM 87502
(505) 827-2944 Fax: (505) 827-2965
mary_ann_menetrey@nmenv.state.nm.us

Mary Ann Menetrey has worked for the Environment Department since 1991 and has held her current position since 2000. Her primary responsibilities include oversight of ground water discharge permitting, including closure plans and abatement actions, for all hard rock mining operations regulated under the New Mexico Water Quality Act and Water Quality Control Commission Regulations. She also worked six years as a project manager and soil scientist for an environmental consulting firm. Mary Ann received her B.S. degree in soil science from California Polytechnic State University and was an M.S. candidate in soil science at the University of California at Davis.

Nordstrom, D. Kirk
Hydrogeochemist, U.S. Geological Survey
Water Resources Discipline, National Research Program
3215 Marine St., Suite E-127
Boulder, CO 80303
(303) 541-3037 Fax: (303) 447-2505
dkn@usgs.gov

Kirk Nordstrom joined the U.S. Geological Survey in 1980 as a project chief for the Trace Element Partitioning Project that has evolved into the Chemical Modeling of Acid Waters Project. His main research has focused on processes affecting water quality from the mining of metals in the western U.S. He has studied pyrite oxidation, reported on acid mine waters having negative pH, developed and applied geochemical models to acid mine waters, studied microbial reactions in acid mine waters, and demonstrated the deleterious consequences of mine plugging. He has also worked on research related to radioactive waste disposal. Nordstrom was an assistant professor at the University of Virginia for four years. He has published over one hundred fifty scientific reports and papers, given hundreds of lectures, and consulted for many state, federal, and foreign government agencies. He holds a B.A. in chemistry from Southern Illinois University, an M.S. in geology from University of Colorado, and a Ph.D. in applied earth sciences from Stanford University.

North, Robert M.
Chief Geologist, Phelps Dodge
Chino Mines Co.
P.O. Box 7
Hurley, NM 88043
(505) 537-3381
rnorth@PhelpsDodge.com

Robert North has been with Phelps Dodge since 1988. His work at Phelps Dodge has focused on exploration, definition, mineral characterization, and computer geologic modeling of porphyry copper deposits in Arizona, New Mexico, and Chile, and sedimentary copper deposits in Zambia and the Democratic Republic of the Congo. He began his working career in 1978 as mineralogist at the New Mexico Bureau of Mines and Mineral Resources. Research and public service projects while at the bureau focused on the geology and mining history of New Mexico's metallic mineral deposits and included a number of publications on the subject. Bob's title was changed to economic geologist/mineralogist in 1986, reflecting his focus on economic geology. Bob completed B.S. and M.S. degrees in geology at Illinois State and Northern Illinois Universities, respectively.

Pfeil, John J.
Geologist, Mining and Minerals Division
Energy, Minerals and Natural Resources Department
1220 South St. Francis Drive
Santa Fe, NM 87505
(505) 476-3407
jpfeil@state.nm.us

John Pfeil is a program manager for Mining and Minerals Division, Energy, Minerals and Natural Resources Department and is responsible for the development of several publications relating to New Mexico's extractive minerals industry. He has worked for a number of private and government organizations whose focus is the management and protection of natural resources. He attended the University of Wisconsin at Madison and graduated with degrees in both geology and history.

Power, Thomas Michael
Professor and Chair, Economics Department, University of Montana
32 Campus Drive #5472
Missoula, MT 59812-5472
(406) 243-2925
tom.power@mso.umt.edu

Thomas Power is professor of economics and chairman of the Economics Department at The University of Montana. He specializes in natural resource and regional economics in which fields he has written five books, most recently Post-cowboy economics: Pay and prosperity in the new American West (Island Press, 2001). He received his Ph.D. from Princeton University.

Price, L. Greer
Senior Geologist/Associate Director
Chief Editor
New Mexico Bureau of Geology and Mineral Resources
New Mexico Institute of Mining and Technology

801 Leroy Place
Socorro, NM 87801
(505) 835-5752 Fax: (505) 835-6333
gprice@gis.nmt.edu
Greer Price currently directs the publications program at the bureau. His experience includes seven years as a geologist working in the oil patch, ten years with the National Park Service, and four years as managing editor at Grand Canyon Association. He has served on the board of the *Publishers Association of the West* and is currently on the board of the *New Mexico Geological Society Foundation*. His focus for many years has been on the interpretation of geology for the general public. His career has involved teaching, writing, and field work throughout North America. He is the author of *An Introduction to Grand Canyon Geology*. He holds a B.A. and an M.A. in geology from Washington University in St. Louis.

Robertson, Andrew MacG.
Robertson GeoConsultants, Inc.
580 Hornby Street, suite 640
Vancouver, BC V6C 3B6
Canada
(604) 684-8072 Fax: (604) 684-8073
rgc@infomine.com
Andrew Robertson has thirty-five years of experience in geotechnical engineering applied to mines. He was a founding partner in 1974 of the international firm of mining and geotechnical engineers, SRK Consultants, and was responsible for the development of their U.S. and Canadian practices. He formed Robertson GeoConsultants Inc. in 1995 and provides an international consulting practice in mining geotechnics, mine waste management, stability and environmental liability assessments for mines, risk management, closure planning and decision analyses for mines. He serves on a number of independent geotechnical review boards for major tailings dams and large risk mining geotechnical structures internationally and has served as a technical advisor to a number of U.S. states and government agencies internationally on these matters. He has written over seventy five technical papers on these topics. He has a B.S. in civil engineering, a Ph.D. in rock mechanics.

Scholle, Peter A.
State Geologist, Director
New Mexico Bureau of Geology and Mineral Resources
New Mexico Institute of Mining and Technology
801 Leroy Place
Socorro, NM 87801
(505) 835-5294 Fax: (505) 835-6333
pscholle@gis.nmt.edu
Peter Scholle has had a rich and diverse career in geology: nine years with the U.S. Geological Survey, four years directly employed by oil companies (plus many additional years of petroleum consulting), seventeen years of teaching at two universities, and now a career in state government at the New Mexico Bureau of Geology and Mineral Resources. His main areas of specialization are carbonate sedimentology and diagenesis as well as exploration for hydrocarbons in carbonate rocks throughout the world. He has worked on projects in nearly twenty countries, with major recent efforts in Greenland, New Zealand, Greece, Qatar, and the Danish and Norwegian areas of the North Sea. A major focus of his studies dealt with understanding the problems of deposition and diagenesis of chalks, a unique group of carbonate rocks that took on great interest after giant oil and gas discoveries in the North Sea. His career has also concentrated on synthesis of sedimentologic knowledge with the publication of several books on carbonate and clastic depositional models and petrographic fabrics. He and his wife have published many CD-ROMs for geology, oceanography, and environmental science instructors, and they currently are developing computer-based instructional modules and expert systems in carbonate petrography. He has been president of Society for Sedimentary Geology and currently is treasurer of the American Geological Institute and president-elect of the Association of American State Geologists. Peter Scholle received a B.S. in geology from Yale University and continued his studies at the University of Munich on Fulbright/DAAD Fellowships and at the University of Texas at Austin. Scholle received M.S. and Ph.D. degrees in geology from Princeton University.

Shaw, Shannon
Geochemist/Mineralogist
Robertson GeoConsultants, Inc.
580 Hornby Street, Suite 640
Vancouver, BC V6C 3B6
Canada
(604) 684-8072 Fax: (604) 684-8073
shannon@infomine.com
Shannon Shaw has worked with Robertson GeoConsultants for more than eight years providing consulting services to the mining industry internationally in fields related to material characterization, acid rock drainage, mine waste management, closure planning, and multi-stakeholder facilitation for options evaluations and risk assessments. She has served as project manager and third party reviewer for many projects and has written a number of technical papers and presentations. She has a B.S. in geological and chemical sciences, and an M.S. in geological sciences with a specialization in sulfide oxidation.

Shepherd, Holland
Mining and Minerals Division
Energy, Minerals and Natural Resources Department
1220 South St. Francis Drive
Santa Fe, NM 87505
(505) 476-3437 Fax: (505) 476-3402
hshepherd@state.nm.us
Holland Shepherd is currently working as the program manager of the New Mexico Mining and Minerals Division, Mining Act Reclamation Program. He has been working for the Mining and Minerals Division for the Coal and Mining Act Programs for about ten years. Before coming to New Mexico he worked for the Division of Oil, Gas and Mining for the state of Utah. Shepherd has worked in the environmental regulatory field for about eighteen years. Shepherd's formal training is in environmental science at the University of Denver and mined land reclamation at Colorado State University.

Shields, Brian
Executive Director, Amigos Bravos
P.O. Box 238
Taos, NM 87571
(505) 758-3874 Fax: (505) 758-7345
bshields@amigosbravos.org
Brian Shields is executive director of Amigos Bravos and Friends of the Wild Rivers, a statewide river advocacy organization with offices in Taos and Albuquerque, New Mexico, which he helped form and has led for over seventeen years. For the past sixteen years Brian has been instrumental in creating a community-based "river's movement" along the Río Grande and in the communities of northern New Mexico. Brian is involved in a number of community collaborative efforts including citizen monitoring, river restoration, water protection, and policy development. Those efforts include working with land-based indigenous communities, irrigators, environmentalists, ranchers, social and environmental justice organizations, and government agencies to restore clean water, natural flows and native fisheries. Brian has taken a leadership role in establishing various community-based groups and organizations including the Río Grande/Río Bravo Basin Coalition, Río Grande Alliance, Alliance for the Río Grande Heritage, Río Chama Coalition, Rio Pueblo/Rio Embudo Watershed Protection Coalition, Reviva el Río Costilla, Red River Watershed Association, the New Mexico Mining Act Network, the Mining Impacts Communications Alliance, New Mexicans for Sustainable Energy and Effective Stewardship, and the Coalition for the Valle Vidal.

Shomaker, John W.
John Shomaker & Associates, Inc.
2703 Broadbent Parkway NE, Suite B
Albuquerque, NM 87107
(505) 345-3407 Fax: (505) 345-9920
jshomaker@shomaker.com

John Shomaker is president of John Shomaker & Associates, an Albuquerque consulting firm specializing in water planning, ground water supply, water quality, and water rights matters. Clients include many of New Mexico's municipalities, investor-owned water utilities, mining and industrial enterprises, and state government. John has provided expert testimony in many New Mexico State Engineer hearings, in Water Quality Control Commission and Environment Department hearings, in State District Court, and in interstate water litigation before a U.S. Supreme Court Special Master. His interests include ground water geology, ground water modeling, well hydraulics, and water-resource planning. He is the author, co-author, or editor of some forty publications and over one hundred twenty-five consulting reports available in the public record. He was with the U.S. Geological Survey, Water Resources Division from 1965 to 1969, the New Mexico Bureau of Mines and Mineral Resources from 1969 to 1973, and has been a consultant since 1973. He holds B.S. and M.S. degrees in geology from the University of New Mexico; an M.A. in the liberal arts from St. John's College, Santa Fe; and an M.Sc. and a Ph.D. in hydrogeology from the University of Birmingham (England).

Smith, Kathleen S.
U.S. Geological Survey
M.S. 964, Denver Federal Center
Denver, CO 80225
(303) 236-5788 Fax: (303) 236-3200
ksmith@usgs.gov
Kathy Smith is research geologist/geochemist with the U.S. Geological Survey in Denver, Colorado. She has worked at the U.S. Geological Survey for twenty-five years and has spent the past fifteen years working at mined sites. Her research interests include water/rock interactions, metal-transport processes, metal speciation, metal-sorption processes, mine-waste geochemistry, and leaching methods. She is currently overseeing an interdisciplinary research project to develop methods and tools for mine-site characterization. Kathy holds a B.S. in environmental geology from Beloit College in Wisconsin, and an M.S. and Ph.D. in geochemistry from the Colorado School of Mines.

Uhl, Mary
Air Quality Bureau, New Mexico Environment Department
2048 Galisteo
Santa Fe, NM 87505
(505) 955-8086
Mary Uhl is manager of the Planning and Policy Section of the Air Quality Bureau of the New Mexico Environment Department. She has seventeen years of experience in the field of environmental science including thirteen years with the Air Quality Bureau. She served as co-chair of the Western States Air Resources Council Technical Committee and has chaired the Western Regional Air Partnership's Modeling Forum for several years. Upon completion of her degree, Uhl worked as a research associate at Purdue, continuing research in the development of meteorological models under funding from the National Aeronautics and Space Administration. Uhl received her B.S. in math and chemistry and completed her M.S. degree in atmospheric sciences at Purdue University.

van Zyl, Dirk
Director, Mining Life-Cycle Center
Mackay School of Earth Sciences and Engineering
Mail Stop 173
University of Nevada, Reno
Reno, NV 89557
(775) 784-2039 Fax: (775) 784-1833
dvanzyl@mines.unr.edu
Dirk Van Zyl is the director of the Mining Life-Cycle Center and professor and chair of Mining Engineering at the Mackay School of Earth Sciences and Engineering of the University of Nevada, Reno. Dirk has about thirty years experience in research, teaching, and consulting in tailings and mine waste rock disposal and heap leach design. Most of his work has been focused on geotechnical and environmental mining engineering aspects to provide solutions for environmental and human health protection. He previously taught at the University of Arizona and Colorado State University. Dirk has more than seventy publications to his credit and is the recipient of the three awards from the Society for Mining, Metallurgy, and Exploration. Dirk received a B.S. degree in civil engineering and a B.S. (Honors) from the University of Pretoria, South Africa. He also received an M.S. and Ph.D. in geotechnical engineering from Purdue University. He also completed an Executive M.B.A. at the University of Colorado.

Wagner, Anne M.
Manager, Environmental and Health Services
Molycorp, Inc.
P.O. Box 469
Questa, NM 87556
(505) 586-7625
awagne@molycorp.com
Anne Wagner has been working for Molycorp at the company's Questa facility for the past eleven years. As manager of environmental and health services, Wagner is responsible for supervising many of the company's large environmental and reclamation projects. She has over twenty years of experience in the environmental sciences including specialization in plant physiology, and she has published several papers on plant propagation and revegetation techniques for high altitude steep slope reclamation. In addition, she is currently chair of the New Mexico Mining Association, Environment Committee. Wagner received her B.S. in biology from Fort Hays State University, in Kansas, and an M.S. and Ph.D. in agronomy and horticulture from New Mexico State University.

Williams, R. David
Geologist, Bureau of Land Management
106 North Parkmont
Butte, MT 59701
(406) 533 7655
david_r_williams@blm.gov
David Williams has worked for the Bureau of Land Management as a geologist since 1977. He is involved with all aspects of federal mining law and mine permitting in southwest Montana. Dave has been active in acid drainage issues since the 1990s and has experience in several areas in the U.S. as well as Western Australia. He has presented papers at past International Conferences on Acid Rock Drainage and is on the organizing committee for the upcoming 2006 International Conference on Acid Rock Drainage. Dave has also been active in the Bureau of Land Management, Abandoned Mined Land Program, recently overseeing the completion of site restoration at an abandoned lead zinc mine near Missoula, Montana. David received his B.S. degree from Bates College and his M.S. from the University of Montana.

Yazzie, Melvin H.
Senior Reclamation Specialist
Navajo Abandoned Mine Lands Reclamation Program
P.O. Box 3605
Shiprock, New Mexico 87420
(505) 368-1224
mhyazzie@frontiernet.net
Melvin Yazzie has been with the Navajo Abandoned Mine Lands Reclamation Program as a reclamation specialist and acting department engineer in variable capacities for over fifteen years. His duties have covered inventory and site assessments, project designs, cost estimating, construction monitoring, closeouts, Geographic Information System database development, in-house training and many administrative and partnering duties. He has a B.S. degree in mining engineering from New Mexico Institute of Mining and Technology.

PHOTO CREDITS

Unless otherwise noted, all of the graphics in this volume were completed by Tom Kaus and Leo Gabaldon from material provided by the authors. The metals & industrial minerals map on page 35 appears courtesy of Maureen Wilks and the New Mexico Geological Society. Special thanks to Virgil Lueth for his conceptualization of the graphic on page 29. Individual photographers hold copyright to their works.

Front cover, i	Ralph Lee Hopkins
5, 6	William Stone
10	William Stone
11	Ralph Lee Hopkins
19 left	Collection of Sandia Software
19 right	U.S. Geological Survey
20	Gary Rasmussen
21	Author's collection
22 top	New Mexico Bureau of Geology Photo Archives, No. H-1021; postcard view courtesy Spencer Wilson.
22 bottom	New Mexico Bureau of Geology Photo Archives
23 top	Author's collection
23 bottom	New Mexico Bureau of Geology Photo Archives, No. SG-432; photo by John Schilling, courtesy Connie Schilling.
26, 27, 28	Molycorp, Inc.
29	Leo Gabaldon, after Virgil Lueth
31	Virginia McLemore
34, 37 top	L. Greer Price
37 bottom	Richard Johns, New Mexico Bureau of Geology Photo Archives
38	Peter Harben
39, 40	William Stone
42, 43, 44	EARTHWORKS
47	Steve Blodgett/LightHawk
48	James Kuipers
52	Mining and Minerals Division, Energy, Minerals and Natural Resources Department
53	Virginia McLemore
60, 61	U.S. Geological Survey; photo on 61 courtesy of philip Hagerman
67	Kirk Nordstrom, U.S. Geological Survey
73, 74	Mining and Minerals Division, Energy, Minerals and Natural Resources Department
80	Paul G. Logsdon, courtesy of Marcia L. Logsdon
81	Gene Darnell, State Land Office
83, 84	Adriel Heisey
118, 119	Mining and Minerals Division, Energy, Minerals and Natural Resources Department
131, 132	Laurence Parent
139	William Langer, U.S. Geological Survey
141	Abe Gundiler
142 top	R. David Williams
142 bottom, 143	Abe Gundiler
145	Mischa Richter, from *The New Yorker*

Back cover:

Clockwise from top left: William Stone, Laurence Parent, William Stone, Steve Blodgett/LightHawk, Adriel Heisey, Molycorp, Inc.

ACRONYMS

AAS	alternative abatement standards
AMD	acid mine drainage
AML	abandoned mine lands
ARD	acid rock drainage
BBER	Bureau of Business and Economic Research
BLM	U.S. Bureau of Land Management
BMP	best management practice
CERCLA	Superfund
DEQ	Montana Department of Environmental Quality
EMB	Environmental Management Bureau (Montana DEQ)
EPA	U.S. Environmental Protection Agency
FDIC	Federal Deposit Insurance Corporation
FMEA	Failure Modes and Effects Analysis computer model
HRI	Hydro Resources, Inc.
ICMM	International Council on Mining and Metals
ISL	in situ leaching
ITP	Industrial Technologies Program (U.S. Department of Energy)
LA	load allocation
LES	Louisiana Energy Services
MAA	Multiple Accounts Analysis computer model
MIW	mining influenced waters or mine impacted waters
MMD	Mining and Minerals Division (New Mexico Energy, Minerals and Natural Resources Department)
MMSD	Mining Mineral and Sustainable Development project
MPC	Montana Power Company
MSHA	Mine Safety and Health Administration
NAMLRP	Navajo Abandoned Mine Lands Reclamation Program
NAS	National Academy of Sciences
NEPA	National Environmental Protection Act
NGO	non-government organization
NPDES	National Pollutant Discharge Elimination System
OSM	Office of Surface Mining
RAMS	Restoration of Abandoned Mine Sites (U.S. Army Corp of Engineers)
R&D	research and development
RFP	request for proposals
SARM	sustainable aggregate resource management
SMCRA	Surface Mining Control and Reclamation Act
SXEW	solvent extraction/electrowinning
TDS	total dissolved solids
TMDL	total maximum daily load
USFS	U.S. Forest Service
USGS	U.S. Geological Survey
WLA	wasteload allocation
WQCC	Water Quality Control Commission

Geologic Time		Geologic Record		Mineral Deposits	Events
		Igneous	Sedimentary		
CENOZOIC — Quaternary	Holocene		Surficial deposits in modern drainages and slopes, wind-blown deposits on plateau.	Sand and gravel deposits are quarried as aggregate for use in construction.	Modern landscape has formed. Rio Grande integrated to Gulf of Mexico. Rio Grande/Red River downcuts into Taos Plateau. Ancestral Rio Grande flows across volcanic plateau from Red River headwaters.
	Pleistocene (1.8)	Jemez volcanic field	Alluvium along Rio Grande gorge, within modern tributary valleys, and alluvial-fan deposits on mountain-front piedmonts.		
CENOZOIC — Tertiary	Pliocene (5.3)	Taos Plateau volcanic field / Servilleta Basalt	SANTA FE GROUP — High gravels and basin-fill deposits, including Blueberry Hill Fm.	Blueberry Hill Formation is quarried for aggregates. Volcanic rocks are quarried for perlite, cinder, road maintenance and construction, etc.	Basalts flow south and southeast over Santa Fe Group sediments. Alluvial fans prograde basinward from Sangre de Cristo Mts.
	Miocene (7.1 / 23.8)	Earliest Jemez Mts. eruption	Tesuque Fm.	Sand and gravel deposits are quarried as aggregate for use in construction. Tertiary volcanic rocks host precious metals (gold, silver, lead, zinc, molybdenum, chrome, etc.)	Earliest Jemez Mts. eruption. Amalia Tuff erupted onto early-rift sediments. Questa caldera forms. Rio Grande rifting begins. Early Santa Fe Group sediments and volcanics in protobasins at onset of extension of rift. Ash-flows and lava flows from San Juan volcanic field. Widespread erosion of highlands.
	Oligocene (28.5)	Latir volcanic field / San Juan volcanic field	Picuris Fm.		
	Eocene (54.8)				Laramide orogeny. Laramide highlands. Much of New Mexico covered by seas that retreat and advance. Coal swamps.
	Paleocene (65)			Sedimentary rocks contain large reserves of coal and coalbed methane.	
MESOZOIC	Cretaceous (144)		Mesozoic rocks are well represented to the east and west of the map area.	Building stone, dinosaur bones.	Streams, lakes, and sand dunes.
	Jurassic (206)				
	Triassic (248)			Sedimentary rocks are used for landscaping and building.	Sediments accumulate in Taos trough. Rivers and streams deposit sediments. Uplift and erosion for 40 million years.
PALEOZOIC	Permian (290)		Sangre de Cristo Formation		
	Pennsylvanian (323)		Alamitos Formation / Flechado Formation		Ancestral Rocky Mountain orogeny. Shallow seas.
	Mississippian (354)		Arroyo Peñasco Group		
	Older Paleozoic Rocks (543)				Sediments were not deposited in northern New Mexico because the region was a landmass that shed sediments into the seas to the south.
PRECAMBRIAN (4,600)		Igneous and metamorphic rocks, 1.4 – 1.75 billion years old	Vadito and Hondo Groups 1.7	Precambrian rocks are mined for clay, mica, copper, gold, rare earth metals, and collectible minerals.	

Millions of years (not to scale)